T0324909

Chocolate Crisis

CHOCOLATE CRISIS

Climate Change and Other Threats to the Future of Cacao

DALE WALTERS

UNIVERSITY OF FLORIDA PRESS

Gainesville

26 25 24 23 22 21   6 5 4 3 2 1

Library of Congress Cataloging-in-Publication Data
Names: Walters, Dale, author.
Title: Chocolate crisis : climate change and other threats to the future of cacao / Dale Walters.
Description: Gainesville : University of Florida Press, 2021. | Includes bibliographical
references and index.
Identifiers: LCCN 2020023711 (print) | LCCN 2020023712 (ebook) | ISBN
9781683401674 (hardback) | ISBN 9781683401940 (pdf)
Subjects: LCSH: Chocolate industry. | Cacao—Climatic factors. | Crops and
climate. | Climatic changes—Social aspects. | Cocoa trade—Forecasting.
Classification: LCC HD9200.A2 W35 2021 (print) | LCC HD9200.A2 (ebook) |
DDC 338.1/7374—dc23
LC record available at https://lccn.loc.gov/2020023711
LC ebook record available at https://lccn.loc.gov/2020023712

University of Florida Press
2046 NE Waldo Road
Suite 2100
Gainesville, FL 32609
http://upress.ufl.edu

UF PRESS

UNIVERSITY
OF FLORIDA

To Beverley,
thank you

When one thinks of the marvelously nourishing and stimulating
virtue of cocoa, and of the exquisite and irresistible dainties prepared from it,
one cannot wonder that the great Linnaeus should have named it
theo broma, "the food of the gods."

BRANDON HEAD,
The Food of the Gods: A Popular Account of Cocoa
[London: R. Brimley Johnson, 1903]

Contents

Figures

Acknowledgments

Despite my lifelong love of cacao, and chocolate, I am not a cacao researcher. I spent 40 years working on diseases of temperate crops. So, I knew that when I embarked on the journey of writing this book, I would need guidance along the way from people who have actually worked on cacao and the diseases and pests that afflict it. The first person who came to mind when I first thought of writing this book (I was walking the dogs on the sand dunes near my home in Scotland at the time) was Harry Evans. I had met Harry back in the 1980s when we tried to get a joint project going, and I knew of his work on cacao diseases even before that, when I was an undergraduate in the mid-1970s. Harry agreed to read some chapters for me, but he actually read the entire manuscript. He provided a great many helpful suggestions and corrected mistakes. I am truly grateful to him. Martijn ten Hoopen of CIRAD and the Cocoa Research Centre of the University of the West Indies (UWI) at St. Augustine in Trinidad, very kindly read the chapters on diseases and pests, and I am grateful to him for his corrections and suggestions. I am also grateful to Frances Bekele of the Cocoa Research Centre at UWI in Trinidad for reading through the early chapters of the book, and for her encouragement. Four anonymous reviewers provided many suggestions and much food for thought, and I am grateful to them for the effort they spent going through my manuscript. Any errors that remain are entirely my responsibility.

Most of the images that appear in the book were kindly provided by Harry Evans, Martijn ten Hoopen, and Ashley Parasram of Trinidad and Tobago Fine Cocoa Company Limited. The image shown in figure 1.2 was provided by Simon Martin of Pennsylvania State University, USA.

When I first approached University of Florida Press (UFP) with the idea for this book, Linda Bathgate was very supportive and I am grateful to her for believing in the book. I am immensely grateful to Meredith Morris-Babb at UFP for seeing me through the revisions and guiding me as the book made its journey toward publication. She has been an immense help to me, dealing with my many questions with good humor and understanding.

Writing this book has been a genuine pleasure, taking me back to my childhood, and providing me the chance to immerse myself in the world of cacao. I have my late father, Ronald Kevin Walters, to thank for nurturing my early interest in cacao. But my greatest thanks go to my long-suffering wife, Beverley, who loves and supports me through all of my writing journeys.

Prologue

I spent my childhood in Trinidad. I was not born in Trinidad, but some 4,400 miles away in the Welsh seaside town of Tenby. How I ended up in Trinidad is an amazing story. My dad, a Tenby boy, was in the merchant navy and while in the Caribbean had to be rushed to the nearest hospital with acute appendicitis. The nearest hospital was in Port-of-Spain in the island of Trinidad, and following the operation to remove his inflamed appendix, he started up a conversation with the young man in the bed next to him. In due course, his new friend's sister came to visit, and it was love at first sight. My dad was not quite 18 and had to return to the UK for his national service. My mum followed him to Tenby, where they were married in 1956 and I was born the following year. After his national service, my parents decided that a life in tropical Trinidad sounded appealing, and when I was three, they set sail for Port-of-Spain. All my childhood memories are of Trinidad, and some of the most vivid involve cacao (*Theobroma cacao*), not the drink, nor for that matter any of the confections made from cacao, but the tree. You see, my dad worked on various plantations in Trinidad, many of which grew cacao.

Although my earliest memories involve watching cartoons (something I still do, despite being 62), one clear memory from my early childhood is of walking through trees in a cacao plantation with my dad. I was probably about 8 years old and at that time we lived in the northeast of Trinidad, in the tiny village of Toco. Situated in the beautiful Northern Range, Toco is the most northeasterly village in Trinidad, and back in the mid-1960s, when our family lived there, it was remote. In fact, up until 1930, there were no roads connecting Toco to the rest of the island; apparently, the main way to reach Toco was using the island steamship service. I remember the seemingly interminable car journeys to get to our house. But the hours spent in the car were worth it. Toco has the most spectacular scenery, with a rocky coastline and sandy beaches where you could play all day without seeing anyone else. Our house was on a hill, surrounded by forest, but with a view northward to the rocky coastline, the crashing waves, and beyond that, the island of Tobago. My dad was working on a cacao estate, and just a short walk from the house was all it took to be surrounded by cacao trees. I remember walking with him into what looked like forest—in fact it was pretty much forest, since cacao trees are grown in the shade of much taller trees, which in Trinidad were mostly Immortelle (*Erythrina poeppigiana*). These giants grow to a height of about 25 meters, and they flower at the end of the rainy season, in December. Immortelle flowers are a brilliant orange, and around Christmas the whole area was a mosaic of dark green leaves and wonderful orange blossom.

If I close my eyes, I can transport myself back to my childhood, back to my eight-year-old self walking through the cacao and Immortelle with my dad. It was a world of shade, of light and dark, and of silence, interrupted only by the songs of birds I could not see. What sticks in my mind are the cacao pods: large, oval-shaped structures, many of which were red, though some were orange or yellow. The weirdest thing about the pods was that they just hung from the main trunk of the tree. That was so unlike any other fruiting tree I had seen— mango (*Mangifera indica*), guava (*Psidium guajava*), lime (*Citrus aurantifolia*), and tamarind (*Tamarindus indica*).

The other vivid memory I have from this time is of cacao beans drying in the sun. Once the seeds were removed from the pods—this was done in Trinidad by splitting the pods in half with a cutlass—they were "sweated" (fermented) in boxes, after which the beans were spread out on large wooden floors to dry in the sun. The smell was so powerful and distinctive. On a visit to my parents in Trinidad in 1990, I went to a cacao estate in Gran Couva and the smell of the drying cacao beans took me straight back to my childhood. As a boy in Toco, I remember workers "dancing the cacao," when the dried beans were polished by dancing on them with bare feet.

Although we moved about quite a bit during my childhood in Trinidad, and I grew up surrounded sometimes by sugar cane and other times by coco- nut trees, it is cacao that holds the dearest memories for me. Cacao also had a profound effect on the course of my life. On our walks through the cacao, my dad would point out the various ailments affecting the cacao. Two of these aroused my boyhood curiosity: black pod and witches' broom. I heard these terms a great deal, and when I was a bit older, in my early teens and becoming interested in science, I found out that these two ailments of cacao were caused by microbes. I was intrigued and marveled at how two organisms I could not see could cause so many problems for cacao. I wanted to know more, to find out how these microbes could cause such damage to the trees and their fruits; I also wanted to be able to do something about it. When the time came for me to go to university, I chose to study plant science. I recall my first day at Wye College, deep in the Kent countryside in the UK, when new first-year students had to move along a table of college staff, completing forms and collecting information on modules and accommodation. One of the last members of staff I encountered that day was the college secretary, Dr R. E. Wyatt. He asked if I had thought about what I might do with my plant science degree, which I thought was assuming a great deal, since this was my first day, after all. I told him I wanted to be a plant pathologist. He looked at me, smiled, and said that before he became an administrator, he had been a plant pathologist. He then asked why I had chosen plant pathology, and I told him I wanted to find a way of controlling black pod and witches' broom on cacao. That interest in cacao and its diseases never left me, so much so that in my final year, not only did I carry out my special study on black pod, but I also carried out another study for a plant virology module, on *Cacao swollen shoot virus*. I have both studies in front of me as I write these words and just looking at their covers takes me back to 1978. I spent the summer of 1977 back at home in Trinidad (by this time we lived in Rio Claro in the southeast of the island), and I obtained a lot of information on black pod from the library at the University of the West Indies at St. Augustine. I got my degree in 1978, but I never did get to work on cacao. Instead, I spent the next 39 years working on diseases of temperate crops, mainly barley. I am now retired, but that interest in cacao has never left me, and my childhood memories of cacao are still powerful.

The diseases that captured my imagination as a boy are still a major threat to cacao production, and other diseases and pests also threaten the crop. Added to this mix of microbial and insect attackers is climate change, which poses a serious threat to cacao, as it does to other crops across the globe. A number of excellent books for the general reader deal with cacao, particularly its history, botany, natural history, and the processing of cacao to make chocolate. None has tackled the threats to cacao from diseases, pests, and climate change. This triumvirate of topics form the core of this book. My intention is to provide an account for the general reader of the agents responsible for damaging cacao: what they are, how they cause damage, how the cacao tree deals with these attacks, and what we can do to help. I also want to look at the threat posed by the changing climate and what it means for the cacao tree, its diseases and pests, and the farmers who grow cacao. However, before we get to that point, we should acquaint ourselves with cacao and its history.

Abbreviations

ASM	acibenzolar-S-methyl
CEPLAC	Comissão Executiva do Plano da Lavoura Cacaueira
CIAT	International Center for Tropical Agriculture
CRISPR	Clustered Regularly Interspaced Short Palindromic Repeats
CRVV	cacao red vein virus
CSSCDV	cacao swollen shoot CD virus
CSSTAV	cacao swollen shoot Togo A virus
CSSV	cacao swollen shoot virus
DNA	deoxyribonucleic acid
ENSO	El Niño–Southern Oscillation
ICCO	The International Cocoa Organisation
ICI	International Cocoa Initiative
ICTA	Imperial College of Tropical Agriculture
MF	morphological form
NASA	National Aeronautics and Space Administration
NEP	Necrosis and Ethylene-inducing Protein
PCR	polymerase chain reaction
PNG	Papua New Guinea
ppm	parts per million
REDD+	Reducing Emissions from Deforestation and Degradation
RNA	ribonucleic acid
STCP	Sustainable Tree Crop Programme
UTZ	Utz Quality Foods, Inc
VSD	vascular streak dieback

Author's Note

The words "cacao" and "cocoa" tend to be used interchangeably, but technically, there is a distinct difference between them. Cacao refers to the raw material: the beans and the tree from which they are harvested. Cocoa is produced once the beans have been roasted and ground. Chocolate is obtained once the cocoa is mixed with milk, sugar, and cocoa butter.

Across the globe, however, cacao and cocoa are used differently. So, in some parts of the world, Africa, southeast Asia, and Britain, for example, cocoa refers to the raw material—the tree, pods, and beans—whereas elsewhere—South, Central, and Latin America, and the United States, for example—the tree, pod, and beans are referred to as cacao.

This can all be terribly confusing, so, in this book, I use *cacao* to refer to the tree and its products prior to roasting, and *cocoa* to refer to the product following roasting and before its processing into chocolate. There are some instances in the book where "cocoa" is used, not in the sense of processed cacao beans, but in the names of organisms and organizations. For example, the cacao pest *Conopomorpha cramerella* is known as the cocoa pod borer in the scientific literature, and various organizations use "cocoa" in their name, including the International Cocoa Gene-banks, and the World Cocoa Foundation.

1

Food of the Gods

First Encounters with Cacao

The first encounter of the Old World with New World cacao was made during Columbus's fourth voyage. He had set sail from Spain on 9 May 1502 and, forbidden from landing on Hispaniola, he headed for Jamaica. He missed Jamaica and eventually ended up at the island of Guanaja, some 70 km off the coast of Honduras. According to the account written by Columbus's second son Ferdinand, two large dugout canoes appeared, carrying not just people, but also cargo consisting of cotton garments, war clubs, small axes, and bells of cast copper (Coe and Coe 2013, 108–109). Among this cargo were provisions of roots and grains, as well as small items the Spaniards called "almonds." These were clearly of importance to the natives, since they always picked them up if any were dropped accidentally. Columbus had no way of knowing that these "almonds" were actually cacao beans.

It seems likely that the natives Columbus encountered that day belonged to the Putún or Chontal Maya, who by this time controlled a coastal trade network stretching from the Yucatán Peninsula to the Gulf of Honduras. Cacao beans were important to the Putún not least because they used them as money. But the history of cacao stretches considerably further back than 1502, and the beans were used not only as currency, but also as a beverage.

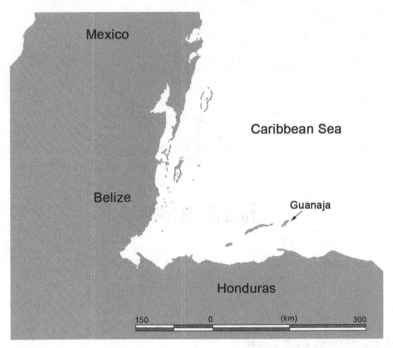

Figure 1.1. Map showing the location of the island of Guanaja, in the Caribbean Sea, off the coast of Honduras.

Cacao before Columbus

The first civilization of the Americas, the Olmecs, flourished from 1500 to 400 BC and was centered on the humid lowlands of the Mexican Gulf coast. The Olmecs spoke an ancestral form of the Mixe-Zoquean family of languages, some of which is still spoken in the lands they once occupied. "Cacao" appears to be a word from that language, originally pronounced *kakawa* (Coe and Coe 2013, 18). It seems possible that the Olmecs were the first to domesticate the cacao plant and to use the beans to make a drink. Evidence to support this suggestion came from the use of modern analytical techniques by chemists at Hershey Foods in the United States. Cacao contains three alkaloids, the most important of which are theobromine and caffeine. In Mesoamerica, cacao is the only plant to contain both alkaloids. When samples were scraped from the inner surfaces of ceramic vessels from the Maya archaeological site at Colha in northern Belize, dated to around 600 BC, the chemists at Hershey found that both theobromine and caffeine were present (Hurst et al. 2002, 289–290). It seemed that people inhabiting Mesoamerica 2,600 years ago were using cacao, probably as a drink.

Two and a half millennia is a long time, but we need to go even further back in time to find the beginnings of chocolate drinking. The Soconusco region, which includes the Pacific coastal plain of Chiapas in southeast Mexico and adjacent Guatemala, is now thought to be where sedentary village culture started in Mesoamerica. Excavations in this area have revealed a culture dating back to 1800–1400 BC. Astonishingly, ceramic drinking jars from these excavations tested positive for the presence of theobromine (Coe and Coe 2013, 36), which suggests that people were drinking chocolate more than 3,800 years ago. That was the state-of-play in late 2017 when I was writing this chapter. Then, a year later, when I was finishing the final chapter, I came across an article published in the journal *Nature,* in which researchers provided evidence for the use of cacao in southeast Ecuador between 5,450 and 5,300 years ago (Zarillo et al. 2018). This is the first directly dated archaeological evidence for the use of cacao in South America and ties in with studies showing that this region was the center of domestication of the cacao tree (see chapter 3). The use of chocolate as a beverage has a much longer history than I had imagined.

Cacao was very important in Mayan culture. So much so that during the Classic period (AD 200–900), one of the last rulers at the ancient city of Tikal in Guatemala (in the Petén region) was called Lord Cacao (Young 2007, 21). Chocolate drinks were used in Mayan betrothal and marriage ceremonies, especially among the rich. At such festivities, people would drink chocolate together—known in K'iche' Maya as *chokola'j,* which is the possible source for the Spanish and English word "chocolate" (Coe and Coe 2013, 61). The Lacandón Maya once ruled a vast domain in the eastern Chiapas, and today just a few hundred survive. They have retained many of their cultural traditions, including the making and use of chocolate drinks for human consumption and as an offering to their gods. To make the drink for their own use, they toasted and ground together fermented and dried cacao beans with toasted corn and then mixed with water containing a foaming agent—a section of a vine called suqir. This mixture was then whipped with a wooden spoon until foam was produced. The foam, apparently the most desirable part, was consumed, after which the remaining liquid was placed on top of the corn and cacao gruel and eaten. The cacao pathologist Harry Evans, whom we will come across later in the book, says that he had this drink in the states of Chiapas and Tabasco in the south of Mexico, where it is called pozol.

Figure 1.2. God L with merchant's pack and cacao tree. Mural detail. Late Classic period. Red Temple, Cacaxtla, Mexico. God L is one of the twin sons of the Maize God, and by causing the death of the Maize God he took possession of the Maize Tree and so came to own cacao and all the wealth it represented. By permission of Simon Martin.

It was known initially as *pochotl,* from the Nahuatl *pozolli* (sparkling), but following the arrival of the Spanish in the early sixteenth century, it became known as *pozol.*

Cacao and chocolate were also important in Aztec culture. They called the cacao tree *cacvaqualhitl,* the pods *cacvacentli,* the beans *cachoatl,* and, apparently, the drink made from the beans was known as *chocolatl* (Coe and Coe, 2013, 61). For the Aztecs, as with the Mayans, chocolate was the preserve of the elite: the royal house, lords and nobility, although warriors were also allowed it, usually supplied in pellet or wafer form. Apparently, a chocolate drink was not taken during a meal, but at the end, much as we might drink port or brandy at the end of a special meal today. As with the Mayans, flavorings were often added to the chocolate drinks. These included vanilla and chili. Chocolate drinks were taken unsweetened, quite unlike the way many of us would drink chocolate today.

Although most books and articles state that the word "chocolate" is derived from the Nahuatl word *chocolatl*, there is no mention of this word in any early source on the Nahuatl language or Aztec culture. In these sources, the word for the chocolate drink is *cacahuatl* or "cacao water." Nevertheless, by the second half of the sixteenth century, the Spaniards were using the word *chocolatl*. This was a drink made up of equal parts cacao beans and ground seeds of the ceiba tree, frothed up with a wooden stick known as a molinillo. It seems that *chocolatl* was transformed into "chocolate" by the Spanish and used by them to describe drinks made from cacao (Coe and Coe 2013, 61).

Cacao beans were used as tributes to powerful rulers in pre-Columbian Mesoamerica. At the time of the Spanish conquest, Aztecs demanded tributes of cacao beans from conquered regions. The tributes often had to be carried great distances to the Aztec capital of Tenochtitlán, which was situated on an island near the western shores of Lake Texcoco in Central Mexico. Cacao was so valuable to the Aztecs that their ruler, Montezuma, had huge storehouses of cacao beans. These beans were treasure and were not used for consumption. In fact, only old and worn cacao beans were used to make their chocolate drink, *chocolatl*.

We know that cacao beans were used as currency, although little information is available on the value of this cacao currency prior to the Spanish conquest. What is available is the purchasing power of cacao beans shortly after the conquest. For example, in Nicaragua, a porter's daily wage was 100 beans. In 1545, he might have purchased a turkey or a rabbit for 100 beans, a turkey egg or an avocado for 3 beans, and a tomato for 1 bean (Coe and Coe 2013, 99–100). As Sophie and Michael Coe say in their excellent book *The True History of Chocolate*, every time an Aztec took a drink of chocolate, he was literally drinking money.

Columbus had no idea of the importance of the cacao beans they acquired from the native canoe in 1502, and it took some time for the Spanish to appreciate the monetary value of cacao. Although they eventually did come to grasp its value to the native peoples as a currency, they found the drink made from the beans not to their liking. It was too bitter. Sugar cane was already being grown in Spain's new territories, having been taken from the Canary Islands to the New World by Columbus in 1493. It wasn't long before cane sugar was being used to sweeten the chocolate drink, making it more palatable to the sweet-toothed Spanish settlers.

Chocolate Comes to Europe

It was another 40 years before chocolate made an appearance in Europe. In 1544, Dominican friars took a delegation of Maya nobles to visit Prince Philip (later to become Philip II) of Spain. The Mayans took with them a great many gifts to be given to the Prince, including receptacles containing beaten chocolate. What is not known is whether Prince Philip sampled the exotic beverage. Although chocolate may have been taken across the Atlantic by monks or nuns moving between Old World and New World monasteries, the first official shipment of cacao beans was sent from Veracruz on the Mexican coast to Seville in Spain in 1585. It wasn't until the first half of the seventeenth century that chocolate became fashionable in the Spanish court. At this time, it appears that the beverage was the same as that being used by the Spaniards of mixed descent living in Mexico. In one recipe from 1644, in addition to cacao beans, nine other ingredients were used, including chili, anise, vanilla, almonds, hazelnuts, and sugar.

Cacao beans were also used to make chocolate confectionery. Apparently, chocolate sweets were being made by nuns in Mexico quite early on and would have been eaten at banquets in Baroque Europe, along with a variety of other sweet delights (Coe and Coe 2013, 134).

Chocolate made its way to England, France, and Italy during the seventeenth century. It appears that the English first came across cacao via the activities of pirates and adventurers who terrorized Spanish ports in the late 1500s. They had no idea what the strange, bitter-tasting beans were, so much so that in 1579, English buccaneers burnt a shipload of cacao beans, thinking they were sheep droppings (Coe and Coe 2013, 161). If only they had realised that the load of beans was worth a fortune! Despite the buccaneer bean-burning episode, the English did come to know about and appreciate cacao. The British captured Jamaica from the Spanish in 1655, and the cacao plantations there became the main source of cacao beans for Britain thereafter. Just a couple of years later, chocolate was being advertised in an English newspaper. By 1660, drinking chocolate was one of the pleasures taken by Samuel Pepys, with chocolate available in coffee houses. Compared with the elaborate methods for making chocolate drinks used by the Spanish and others, the English seemed to want their chocolate in a hurry. One method, described by Philippe Dufour in 1685, involved mixing cacao powder with sugar, adding some hot water, pouring several times from one vessel to another to generate the foam or froth, and hey presto! Chocolate to go!

Chocolate consumption in Europe increased greatly during the seventeenth and eighteenth centuries. During the reign of Charles III of Spain (1759–1788), chocolate consumption in Madrid was said to be in the region of 5.4 million kg every year. This chocolate was strictly for the upper and middle classes, who took it for breakfast after drinking a glass of cold water (Coe and Coe 2013, 207). In the second half of the eighteenth century, coffee houses were the rage in major European cities such as Madrid, providing not just coffee, but also chocolate and tea, for men. Women had to make do with having their beverage brought out to them in their coach.

In London, coffee and chocolate houses were frequented by the nobility, gentry, and burgeoning middle classes. In time, some of these coffee and chocolate houses metamorphosed into gentleman's clubs. One of the best known, White's, started as a chocolate house in 1697. It was run by Francis White, and customers could indulge in a spot of gambling, as well as other more strenuous activities, in addition to imbibing chocolate. Francis White died in 1711 and the establishment was taken over by his widow, who proceeded to lift it up the social ladder (Morton and Morton 1986, 21–23). Even so, White's was still known as "the most fashionable hell in London," a hotbed of decadence, depravity, and destruction, fueled by thick, spicy chocolate (Green 2017). As with other establishments of this type, White's was a place for political decision makers to meet, and its membership has included the British prime ministers Sir Robert Walpole and Sir Robert Peel. Politicians were not the only members of society to frequent these establishments. White's membership has included the writers Evelyn Waugh and Graham Greene. White's is still around but has metamorphosed into a super-exclusive private members' club, with 500 members and a nine-year waiting list, and has lost its association with chocolate (Green 2017).

Chocolate for the Masses

Up until the early nineteenth century, chocolate was a thick and frothy drink made using a method that had not altered much since the Spanish conquest. All that changed in 1828, when a Dutch chemist, Coenraad Johannes Van Houten, patented a process for manufacturing powdered cacao with a very low fat content. With his hydraulic press, Van Houten managed to reduce the fat content of ground cacao beans from 53% to around 27%. The resulting "cake" could be ground to a fine powder. By treating this powder with alkaline salts such as potassium carbonate, the powder's miscibility with water was greatly improved. The days of the protracted method of making the thick, foamy drink were gone. Chocolate could now be prepared with greater ease and was more easily digested once consumed. The large-scale manufacture of cheap chocolate was now possible.

In the mid-1700s, Dr. Joseph Fry gave up his medical practice in Bristol to start a business—the firm of J.S. Fry & Sons, chocolate manufacturers. Following his death, the business was continued by his widow and his son, Joseph Storrs Fry. Mechanization came to chocolate manufacture in 1789, when the son bought a steam engine to grind his cacao beans. Then, in 1847, the Fry firm found a means of mixing a blend of cacao powder with sugar and melted cacao butter instead of cold water, which produced a less viscous paste, capable of being cast into a mould: the first chocolate bars had been produced. This was the world's first true eating chocolate (Coe and Coe 2013, 241).

But Fry had a competitor—John Cadbury. In 1824, he opened a coffee and tea shop in Birmingham, selling chocolate to drink. In 1868, John Cadbury's son George used a model of Van Houten's hydraulic machine to produce cacao powder, which he called Cadbury's Cocoa Essence. This was a great success and was followed a couple of years later by the first "chocolate box"—a selection of chocolate sweets. Chocolate had become big business, not just in Britain, but also in continental Europe and the United States.

Chocolate was indeed big business and was in great demand by the increasingly chocolate-loving public. As a result, it was not immune from the adulteration of food that prevailed in Victorian times. In early nineteenth-century France, potato starch was commonly added to chocolate, although pulverized cacao shells, gum, ground brick, red lead, and vermilion were also used to adulterate chocolate. Some of these (red lead and vermilion) were poisons. Clearly, something had to be done, and eventually, in 1850, a health commission was created for the analysis of foods in Britain (Coe and Coe 2013, 244). Investigations into the adulteration of foodstuffs, including chocolate, inspired the passing of the Food and Drug Act of 1860 and the Adulteration of Food Act of 1872. Neither of these was particularly successful, but they provided the impetus for greater legislative progress. The Sale of Food and Drugs Act of 1875 went further and became the foundation of modern food law. This all had an impact on Britain's chocolate manufacturers. Cadbury admitted to adulterating their chocolate with starch and flour, and subsequently suggested that chocolate manufacturers should state the exact percentage of ingredients on the wrappings of their products. Admitting they had adulterated their chocolate had no lasting impact on Cadbury, because by 1897, they had surpassed Fry

& Sons in terms of total sales of chocolate; Fry & Sons never recovered (Coe and Coe 2013, 245).

Cadbury, and Fry & Sons, were not alone in manufacturing chocolate on a large scale. In continental Europe, the Swiss, famous today the world over for quality chocolate, came to the chocolate business quite late. The first Swiss chocolate factory was opened by François Louis Cailler in Corsier on the shores of Lake Geneva in 1819. By 1826, Philippe Suchard was making chocolate, and today, Suchard is still a major player in the chocolate business. Collaboration between the Swiss chemist Henri Nestlé and the chocolate manufacturer Daniel Peter gave the world the first true milk chocolate in 1879. They used Nestlé's discovery of a way to make powdered milk to produce the first milk chocolate bar. Also in 1879, Rodolphe Lindt invented the process of "conching," a process that greatly improved the quality of chocolate. This was achieved by 72 hours or more of treatment designed to reduce the size of particles in the finished product, giving a high degree of smoothness. Then came the famous "Toblerone," the iconic triangular chocolate bar. It was created by Theodor Tobler and his cousin, Emil Baumann, in 1908, using a special recipe (Toblerone 2019). You can get some idea of the ingredients involved when you realise that Toblerone is an amalgam of "Tobler" and the Italian word for honey and almond nougat, *torrone*.

Across the Atlantic, Milton S. Hershey began making chocolate coatings for his caramel confectionery in 1893, using machinery he saw at an exposition in Chicago. Subsequently, he sold his caramel business, bought a farm in Pennsylvania, and built a chocolate factory on it. In time, this gave rise to "Hershey, The Chocolate Town," a vast development with two schools, gardens, a zoo, and a golf course. Hershey was a considerably larger development than Cadbury's model town of Bournville in Birmingham in the UK. In the 1920s, 23,000 kg of cacao was being produced by Hershey, and in the 1980s, some 25 million Hershey Kisses, small, bite-sized milk chocolates, were being produced daily. Hershey had truly become a pioneer in the mass production of chocolate.

Chocolate: A Global Business

Today, chocolate is a massive global business, worth billions of dollars annually. In 2018, the market leader was Mars Inc., based in the United States, with net sales of some US$18 billion. Cadbury is now part of Mondelēz International, a US-based company that posted net sales in 2018 of US$11.8 billion. Hershey netted sales of US$7.7 billion in 2018, while sales for Nestlé SA of

Switzerland were a bit lower at a still very respectable US$6.1 billion (ICCO 2019). Although sales of chocolate slowed down in the United States in 2017, due in part to a trend toward healthier snacking, sales in the world's largest confectionery market, Western Europe, grew by 2.1% (Candy Industry 2017).

But this multibillion-dollar industry is under threat. The cacao tree is susceptible to the ravages of diseases and pests which cause serious problems for growers. Diseases and pests must be taken seriously since they have a track record of devastating cacao crops. Witness the impact of witches' broom, a disease caused by a fungus, on cacao production in the Bahia region of Brazil in 1988: production there was reduced by 80%, forcing cocoa-growing families to leave their farms for the city (Schmitz and Shapiro 2015). In Southeast Asia, cacao pod borer has caused losses of US$600 million per year, while in Africa, black pod and the cacao swollen shoot virus threaten cacao production in Ghana and Ivory Coast. Then there is climate change. Researchers have suggested that large parts of cacao-producing areas in West Africa will become unsuitable for production of the crop in the future (Schroth et al. 2016). This is a problem not just for the multinational companies that produce chocolate, but also, and crucially, for the farmers who grow cacao in the various regions across the humid tropics.

Like many millions of people across the world, I enjoy chocolate. Although I'm not a chocaholic (at least, I don't think I am), I eat chocolate every day— just a couple of small squares of dark chocolate in the evening. I do this because I love the taste of chocolate and the wonderful smooth, velvety texture that leaves you wanting more. But, to my continual amazement, I restrain myself— everything in moderation, and all that—despite the evidence that chocolate might actually be good for you. But can that be true? Might chocolate actually provide health benefits?

2

Chocolate Can Be Good for You

In 1529, a 30-year-old Franciscan missionary, Fray Bernardino de Sahagún, arrived in what was then known as New Spain (now Mexico). The long-lived Sahagún—he was 90 when he died—spent more than 50 years studying the beliefs, culture, and history of the Aztecs. He helped to establish the first European school of higher education in the Americas, the Colegio Imperial de Santa Cruz de Tiatelolco, in what is now Mexico City. Starting in the 1540s, he used methods that scholars consider a precursor of modern anthropological field techniques, pioneering new methods for gathering ethnographic information and validating its accuracy (León-Portilla 2002). Sahagún was clearly highly organized. In a surprisingly modern approach to obtaining information, he got the elders of various towns in central Mexico to provide answers to questions concerning their culture and religion, which were recorded in the pictorial form of writing used by the native peoples. He then enlisted the help of present and former students of the Colegio to interpret the pictorial answers. The Aztecs spoke Nahuatl, a Uto-Aztecan language still spoken by about 1.5 million people in Mexico. Sahagún got the students to phonetically transcribe the pictorial descriptions into Nahuatl using Latin letters, after which he reviewed the text (he had learned Nahuatl) and provided a Spanish translation. This endeavor took 30 years and was eventually completed in 1575–1577, when Sahagún was in his mid-70s! The result was 12 volumes of what became known as the Florentine Codex, which included a huge list of nearly 300 medical uses for chocolate (León-Portilla 2002; Coe and Coe 2013, 65).

Chocolate in Mayan and Aztec Medicine

In both Mayan and Aztec cultures, cacao-based beverages were drunk by various individuals in society, including priests, high chiefs, and warriors who had distinguished themselves in battle. In Mayan culture, healers belonged to a hierarchy of priests who inherited their position in society. These healers received an education which prepared them to practice medicine by observation, divination, and interpretation of omens (Wilson and Hurst 2012, 35). Mayan cultural heritage was recorded in hieroglyphic writing, a visually striking writing system composed of hundreds of unique signs or "glyphs." These take the form of animals, humans, supernatural beings, objects, and abstract designs. The glyphs express either meaning or sound and were used to write words, phrases, and sentences. Mayan hieroglyphic writing is complex, so much so that it was once considered an unsolvable enigma. Thanks to a major breakthrough by a Russian epigrapher in the 1950s, the phonetic part of the script was decoded, allowing most of the available Mayan texts to be read. Some scripts depict Mayan sacred rituals, with seated gods holding cacao pods, or dishes piled high with cacao beans. Mayan healers were trained to read these hieroglyphs, and the surviving text of a priest from this period includes chants and incantations offered over people suffering from a variety of ailments, including fevers, seizures, and skin disorders. During the healing ritual, the patient was given a mixture of chocolate mixed with peppers, honey, and tobacco juice to drink. In fact, cacao was often mixed with other plant-based remedies to increase the efficacy of the mixture and to make the remedies more palatable. Plants used in conjunction with cacao included Vanilla orchid (*Vanilla planifolia*), Hummingbird plant (*Calliandra anomala*) and Malabar chestnut (*Pachira insignis*) (Wilson and Hurst 2012, 36).

Aztec medicine used a combination of religion, magic, and botany. The Aztecs believed that illness had different causes: punishment from the gods, magic used by their enemies against them, and natural causes. The sick were often tended by priests, but for serious illnesses, a cure doctor or *ticitl* was called for. These physicians used both magic and herbal remedies to treat the illness. The Aztecs were masters of herbal medicine, with an unsurpassed understanding of plants and their uses and access to a vast pharmacopeia. Book 11 of the Florentine Codex contains a large chapter dealing with herbs, including medicinal herbs, which covers 142 distinct species, complete with botanical descriptions as well as details of habitat and the various uses of the plant (Mexicolore, n.d.).

The cacao tree and chocolate were part of Aztec medicine. In Aztec culture, associations were made between chocolate and the heart, and between chocolate and blood. Chocolate beverages were made blood-red by mixing with annatto, a dye made from the seed covering (aril) of achiote (*Bixa orellana*), the fruit of which are heart-shaped and blood-red, while cocoa was commonly mixed with "heart flower" (*Talauma mexicana*) as a remedy for disorders of the heart (Wilson and Hurst 2012, 36). Other Mesoamerican cultures also used cacao for medicinal purposes: to promote the production of breast milk; the pulp in the cacao pods was used to aid childbirth; and the bark of the cacao tree was used in drinks given to women once the baby had been delivered (Wilson and Hurst 2012, 36–37). Looking beyond Mesoamerica, medicinal uses for cacao can be found in other cultures. So, in Colombia, cacao leaves were used in an infusion to produce cardiotonic effects, while in Venezuela, the leaves were used to treat burns, cracked lips, as well as irritations of the genitals and rectum. Today, many of these ancient remedies are still in use. A good example is the use of an infusion of cacao leaves as a cardiac stimulant, which is still practised by the indigenous population of Colombia.

Chocolate as Food and Medicine Following the Spanish Conquest

The initial Spanish dislike of the chocolate drink consumed by the native populations changed once they could sweeten the beverage with sugar. The drink was thought to aid digestion, build strength, and sustain and fatten the individual. While this was so for chocolate made using roasted beans, the same could not be said of the drink made with green or unroasted cacao beans. The latter drink was harmful to digestion, obstructing the liver, spleen, and bowels, leading to melancholy, irregular heartbeats, and shortness of breath (Coe and Coe 2013, 123–124). Sweetening the chocolate drink was not the only change made by the Spanish, because whereas the New World natives took their chocolate cold, the Spaniards liked theirs hot. Although to us this might appear trivial, it was certainly not trivial to the Spanish. European medicine at the time was based on the humoral theory of disease and nutrition that originated in Classical Greece with the writings of Hippocrates (460–377 BC) and which were championed in the later works of Galen (AD 130–210), the renowned physician, philosopher, and writer of ancient Rome.

According to Hippocrates, the human body contains four humors: blood, phlegm, yellow bile, and black bile. Health results from the correct proportions of these humors, whereas an imbalance causes disease. Galen expanded this theory with the addendum that the humors, diseases, and the medicines used to cure disease could also be "hot" or "cold" and "dry" or "moist." Each humor resided in a particular organ in the body and the function of that organ was to produce its humor. This, in turn, determined a person's temperament.

It is important to remember that this thinking was in vogue well before the English physician and natural historian William Harvey published his work on the motion of the heart and blood in living beings in 1628. Harvey's work showed that the heart propelled the blood in a circular course through the body. Harvey's work was not immediately popular. Even though he was supported by the Royal College of Physicians, many found it hard to accept his findings, since they contradicted the theories underlying bloodletting, which was widely used in medical practice at the time. Humoral theory persisted for some time after Harvey's discovery and was still a major influence on medical teaching and practice well into the nineteenth century.

The main principle of Galen's work as a physician was curing by contraries; for example, a "hot" fever should be treated using a "cold" drug. In Galen's scheme, because nutrition formed part of health, food was also subject to this classification. In Europe in the sixteenth and seventeenth centuries, any medicinal benefits of chocolate were judged in relation to these humoral qualities. So, plants were also classed as "hot" or "cold" and "wet" or "dry." In 1570, Philip II sent his physician, Francisco Hernández, across the Atlantic in search of plants with medicinal value. Hernández spent the period 1572–1577 in Mexico, producing his great work—a description of more than 3,000 plant species, along with their Nahuatl names and illustrations by native artists. Unsurprisingly, he classified the plants of Mexico as either hot or cold and wet or dry. But this great work said nothing of how the Aztecs classified these plants or how they used them in medicine. As it happens, the Aztecs also believed that plants were hot or cold and used appropriate plants to correct excess heat or cold in a person's body. For the Aztecs, excess cold in the body was related to retention of water, and illnesses they considered cold or watery, such as gout, were treated by administering a hot herb. Many of these "hot" herbs are diuretics, and giving them to patients suffering from water retention makes sense. However, there was a world of difference between the Aztecs, who understood the herbal remedies they used, and the Europeans of the time, whose slavish adherence to humoral theory was not based on such sophisticated understanding of nature (Coe and Coe 2013, 121–124; Wilson and Hurst 2012, 49–51).

According to Hernández, the cacao bean was viewed as possessing cold and dry qualities and should be used to treat disorders with opposite humoral qualities. Following this line of thinking, drinks made from cacao were used to treat ailments characterized by their hot and wet qualities, such as fevers. For the Aztecs, fevers were not necessarily "hot" and instead of treating feverish patients with a "cold" medicine, hot treatments were often applied, thereby inducing sweating, a practice used in modern medicine. In European medicine, it was also suggested that, being cold and dry, cacao was likely to produce melancholy, a humoral state caused by an excessive concentration of black bile. However, others thought that providing cacao was mixed with other ingredients, it could be taken by people of all temperaments. So, melancholics, with a temperament characterized as cold and dry with black bile, were advised to take a chocolate drink made without chilies but containing some anise seeds. In contrast, those of a phlegmatic temperament, characterized as cold and wet with phlegm predominant, were told to drink their chocolate hot and spicy. Chocolate drinks prepared using hot spices such as cinnamon, anise seed, and black pepper were thought to "warm the stomach, perfume the breath, combat poisons, and alleviate intestinal pain," in addition to "exciting venereal passion." For some, the latter effect condemned cacao, which was thought to be a "violent inflamer of passions" (Fuller 1994, 21). This puts a whole new spin on drinking a cup of chocolate before bedtime.

The first book to be devoted entirely to chocolate was written by the Spanish physician Antonio Colmenero de Ledesma and published in 1631. Titled *Curioso Tratado de la Naturaleza y Calidad del Chocolate* (A curious treatise of the nature and quality of chocolate), it was translated into English, French, Latin, and Italian. According to the author, chocolate benefits health by the "wise and moderate use whereof health is preserved, sicknesse diverted, and cured, especially the plague of the guts; vulgarly called the new disease; fluxes, consumptions, and coughs of the lungs, with sundry other desperate diseases; by it also, conception is caused, the birth hastened and facilitated, beauty gain'd and continued" (Borg and Siegel 2009, 929–942). The medicinal benefits of chocolate were also promoted by English physicians, including Henry Stubbe, who in 1661, was appointed His Majesty's Physician for Jamaica. Just a year later, he produced a book, *The Indian Nectar; or A Discourse Concerning Chocolata*. The idea behind this book was to dispel misconceptions held by the English regarding the use of chocolate as a medicine. Stubbe used case histories from people living in areas where cacao was grown, including evidence from reputable physicians practicing in the New World. The evidence gathered by Stubbe supported the view that chocolate was beneficial to health, and in his book, he described the precise blends of chocolate mixed with other ingredients that were found to be helpful in treating particular ailments (Stubbe 1662).

There was no shortage of information on chocolate for lovers of the beverage in late seventeenth-century England. In addition to Stubbe's *The Indian Nectar,* there was a detailed treatise on cacao and chocolate, written by William Hughes and published in London in 1672. This volume had the snappy title *The American Physician or a Treatise of the Roots, Plants, Trees, Shrubs, Fruit, Herbs Growing in the English Plantations in America: Describing the Place, Time, Names, Kindes, Temperature, Vertues and Uses of them, either for Diet, Physick, and Whereunto is added a Discourse of the Cacao-nut tree, and the use of its Fruit: with all the ways of making chocolate* (Hughes 1672). Hughes had worked aboard a ship that voyaged to the West Indies, and he gained knowledge of the plants of the region and their uses. It seems that Hughes returned to England from the West Indies around 1652 and worked for the dowager Viscountess Conway at Ragley in Warwickshire, probably as a gardener. He compiled his work from his first-hand experience in the West Indies, and although it contained descriptions of plants ranging from potatoes and maize to watermelons and avocados, it also contained a section on chocolate. Unlike earlier publications which provided recipes for making chocolate, the cacao section of his treatise provided recipes and information on cacao production, as well as the preparation of the beans for use in chocolate recipes. The book also contained information on the medicinal uses of chocolate, and Hughes described how people on the American plantations were observed to live for several months on cacao alone, made into a paste with sugar and then dissolved in water. He also recounted how an English Dominican friar, Thomas Gage, who spent time in the New World with Hernán Cortés, drank chocolate two or three times every day for 12 years and was never ill, although he did become "very fat." This side effect of drinking chocolate was also noted by other authors of the time. In his popular and widely translated work on chocolate, published in 1671, Philippe Dufour observed that people who drank a lot of chocolate grew fatter with time (Dufour 1671).

Chocolate as Medicine in Europe

From the late seventeenth century onward, the focus of information on chocolate started to move away from sources based on anecdote and limited case studies, toward detailed analysis of the various concoctions made using chocolate. The pharmacopoeias and formularies of the period were written for professional use and one of the most popular was Nicolas Lémery's *Traité Universel des Drogues Simples,* published in 1698 and translated later into English.

As one would expect, the writing was guided by the medical wisdom of the time, namely the importance of the four bodily humors. According to Lémery "Chocolate agrees, especially in cold weather, with old people, with cold and phlegmatic Persons, and those that cannot easily digest their food, because of the Weakness and Nicety of their Stomachs; but young people of a hot and bilious Constitution, whose humors are already too much in Motion, ought to abstain from it, or use it very moderately" (Lémery 1745).

Across Europe, chocolate was recommended as a remedy for a wide range of medical problems, including elimination of kidney stones, purging the gut, and enlivening the gut (Wilson and Hurst 2012, 72). Charles Darwin's grandfather, the polymath and inventor Erasmus Darwin, promoted the health benefits of chocolate, based on its use in treating gout, while the Swedish botanist, zoologist, and physician Carl Linnaeus, responsible for naming the chocolate tree as *Theobroma cacao,* promoted the use of chocolate for wasting of the body from diseases of the lungs or muscles, hypochondria, and hemorrhoids (Wilson and Hurst 2012, 75).

Chocolate was becoming a widely traveled commodity. Following its discovery by the Spanish in the Americas, it was taken across the Atlantic to Europe and then, in the seventeenth century, was taken back to the Americas, but this time to North America, by the Dutch. By 1712, chocolate was being advertised by a Boston apothecary in his store (Wilson and Hurst 2012, 78). It appeared regularly on prescriptions written in the eighteenth century for the prevention and care of smallpox, and it was included in Benjamin Franklin's *Poor Richard's Almanack* of 1761 as a beverage to be taken by those affected with smallpox.

In the early 1700s, chocolate became linked to milk. The London physician and president of the Royal Society, Sir Hans Sloane, found that adding milk to his chocolate concoction improved its digestibility (Wilson and Hurst 2012, 84). Indeed, he recommended his milk chocolate drink as a restorative. Sloane's milk chocolate recipe was purchased by John Cadbury and Sons and was prepared and marketed as a nourishing, healthy alternative to alcohol. To increase its popularity, Cadbury Brothers advertised their milk chocolate between 1849 and 1885 as a "health food." Chocolate was becoming both a medicine and a food, so much so that in the 1834 *Dispensatory of the United States,* cocoa was recommended not only "as a drink served at morning and evening meals," but also as "an excellent substitute for coffee in dyspeptic cases, being both nutritive and digestible, without exercising any narcotic or other injurious influence" (Wood and Bache 1834).

Helping the Medicine Go Down

Most of us will have experienced the unpleasant taste of some medicines. In fact, the harsh taste of many drugs can put people right off taking their medications. Swallowing a bitter pill will be much easier if it is coated with something with a pleasing taste, usually something sweet, such as chocolate. Indeed, chocolate has long been used as a flavoring to mask the awful taste of various medicines. In the nineteenth century, cocoa powder was used to mask the taste of ingredients ranging from arsenious acid to quinine. Chocolate-flavored medications were also aimed at children, including its use in treatments against intestinal worms. The Penny Chocolate Worm Cakes produced by Robert Gibson and Sons of Manchester and London contained calomel, jalap, and santonin, all ingredients used toward the end of the nineteenth century as a cure against roundworm, thoroughly mixed with chocolate (Apothecaryads 2014). More recently, chocolate has been used to flavor senna-based laxatives and as an agent to tone down the irritant action of other commercial laxatives. As Wilson and Hurst note in their 2012 book *Chocolate as Medicine,* one consequence of using chocolate to flavor laxatives is their overuse and abuse.

During the nineteenth century, chocolate was also used in the naming of products as a means of boosting sales. The discerning individual could buy tonic chocolate, tar chocolate, headache chocolate (presumably to relieve headaches), and even digitalis chocolate for those with heart problems. Those suffering with stomach complaints could choose from a range of products, including digestive chocolate of Vichy salts and stomachic chocolate of cinchona. The world's first digestive biscuit was created in 1892 by Alexander Grant, working at the McVitie & Price biscuit company in Edinburgh. The biscuit was so named because its high baking soda content was thought to aid food digestion. Chocolate came on the scene a bit later, with the creation of the McVitie's Chocolate Homewheat Digestive in 1925. The enduring popularity of this biscuit is reflected in the fact that some 71 million packets of McVitie's chocolate digestives are sold in the UK every year (McVitie's, n.d.).

Chocolate and Human Health in the Modern Era

By the middle of the twentieth century, basing the supposed health benefits of chocolate on anecdote was no longer good enough. A new approach was being used, one which sought experimental evidence to support any claims of medical benefits. As the twentieth century gave way to the twenty-first, claims for medical benefits of chocolate were supported by a flourishing "science" of chocolate (Wilson and Hurst 2012, 134). This meant that biomedical researchers began to use experiments to assess the effects of chocolate on medical disorders.

Chocolate and Cardiovascular Health

Cacao is a rich source of phenolic compounds, which account for up to 18% of the total dry weight of cacao nibs (Lamuela-Raventós et al. 2005). Just how much polyphenol is present in cacao nibs depends on the variety of cacao and the methods used to process the beans. The major polyphenols found in cacao are flavanols such as epicatechin, although small amounts of other polyphenols, such as luteolin and quercitin, are also present. Cacao also contains small quantities of theobromine and caffeine.

The protective effects of cacao flavanols on cardiovascular health have been examined in a great many studies over the past few decades. Taken together, these investigations indicate that consumption of cacao can reduce the incidence of cardiovascular disease (Martín and Ramos 2017). A team of researchers in Switzerland and Germany reviewed 13 studies on the clinical effects of chocolate consumption carried out between 2010 and 2016 and found strong evidence that cocoa consumption increased blood flow (Vlachojannis et al. 2016). These workers also analysed the effects of polyphenols in cocoa on blood flow and found that cocoa products containing around 100 mg of epicatechin reliably increased blood flow. This is great news until you realize that to achieve a dose of 100 mg of epicatechin, you would need to eat between 50 g and 200 g of chocolate. As much as I love chocolate, even I would not eat 200 g of it every day. Too much of a good thing and all that.

But you don't need to consume 200 g daily to gain benefits. Some research has shown that consumption of considerably less can be beneficial. Using magnetic resonance imaging (MRI), researchers at the University of Nottingham in the UK found that consuming a chocolate beverage containing the same amount of flavonoids as 20 g of dark chocolate for five days improved oxygenation of the blood in the brains of healthy young adults, most likely through increasing blood flow (Francis et al. 2006; Wilson and Hurst 2012, 143). More recently, research published by Australian workers showed that consumption of dark chocolate enhanced vascular function and increased cerebrovascular efficiency in postmenopausal women, effects which were probably due to the improved blood flow they measured in the patients (Marsh et al. 2017).

What are such changes likely to mean for people consuming dark chocolate? Scientists in Cambridge and Bogotá reviewed results reported in more than 4,500 research papers, with data gathered from 114,009 people, assessing the effects of chocolate consumption on the risk of developing cardiovascular disease. The results indicated high chocolate consumption was associated with reductions of 37% in cardiovascular disease and 29% in stroke (Buitrago-Lopez et al. 2011). Similar results were obtained in a more recent review. Here, researchers examined 14 studies involving a total of 508,705 people and found that chocolate consumption was associated with decreased risks of developing cardiovascular disease, stroke, and diabetes (Yuan et al. 2017).

So how does chocolate consumption improve blood flow? Increased blood flow occurs when blood vessels dilate, a process known as vasodilation. This happens when smooth muscle cells in the walls of the blood vessels relax, usually in response to a signal from the inner lining of the blood vessels (the endothelium). The signal responsible for this effect is nitric oxide, and it is thought that the flavanols in chocolate protect this process, thereby facilitating increased blood flow. As we get older our arteries stiffen. Stiffening of arteries can also occur when the endothelium is damaged, for example by high blood pressure, smoking, and high cholesterol, a phenomenon known as atherosclerosis. There is evidence that the flavanols in cacao can provide protection against atherosclerosis by reducing levels of bad (LDL or low-density lipoprotein) cholesterol (Martín and Ramos 2017).

Although there is strong evidence that consumption of dark chocolate can increase blood flow and reduce the risk of developing cardiovascular disease, the impact on common measures such as blood pressure is not at all clear. For example, in the study by Vlachojannis and colleagues (2016) mentioned above, there were no consistent effects of chocolate consumption on blood pressure.

Other Medical Effects of Eating Chocolate

There has been much interest in the possibility that cacao consumption might decrease the incidence of various cancers. Although some studies have found that cocoa intake can reduce cancer incidence, other studies found no correlation between cocoa, chocolate, or flavanols and the prevalence of cancers (Martín and Ramos 2017). Another area that has attracted attention is the possible link between cocoa flavanols and the prevention of age-associated neurodegenerative diseases. Again, although some research found a positive effect of consumption of cocoa flavanols or chocolate on cognitive function, other studies could find no effect (Martín and Ramos 2017).

There is a vast and growing body of literature on the medical effects of chocolate consumption and, as made clear in the brief examination above, there is considerable variability in the results obtained. There are several reasons for this. First, there is enormous variation in the amount of chocolate or cocoa product people consumed in the studies. Second, not all studies used chocolate; many used cocoa powder, which has a different composition, and many studies used flavanols. Finally, even in studies where dark chocolate was used, there were large variations in levels of the major polyphenol epicatechin (Wong 2017, 208–210). Some researchers have gone as far as proposing a checklist as a guide to planning research involving cocoa and chocolate that would take these and other concerns into account. Wilson and Hurst suggest that future clinical studies should use amounts of commercially available cocoa that can be adapted easily into diets. This will help researchers better understand the potential benefits of cocoa to human nutrition and health (Wilson and Hurst 2012, 146).

For those looking for an excuse to eat chocolate, finding out that it might be a source of an important vitamin is likely to lead to great joy and, no doubt, extra chocolate chomping. Well, thanks to a group of German researchers, cocoa and chocolate can now be considered sources of vitamin D_2. It appears that cacao beans are susceptible to fungal contamination and as a result often contain large amounts of ergosterol, a sterol abundant in fungi and that also happens to be a precursor of vitamin D_2. Because exposure to ultraviolet light results in the conversion of ergosterol to vitamin D_2, the researchers reckoned that when fermented cacao beans are dried in the sun, the ergosterol in the beans might be converted to vitamin D_2. Their collective hunch was correct, for they found high vitamin D_2 content in cocoa powder and cocoa butter, while among the various chocolates, vitamin D_2 content was highest in dark chocolate and lowest in white chocolate (Kühn et al. 2018). According to the researchers, cocoa and chocolate are clearly dietary sources of vitamin D_2 and food composition databases need to be updated.

Perhaps the news that chocolate is a source of vitamin D_2 has put you in a good mood. But what about eating chocolate? Can that change your mood? Some authors think so and for many, chocolate is considered a "food for moods" (Wong and Lua 2011). This should not be surprising, because cacao is, after all, "the food of the gods." But is there any robust evidence to support the claim that cocoa and chocolate can alter someone's mood?

There is much evidence that flavanols in cocoa and chocolate are involved in cognition-enhancing effects, although this might not be restricted to cocoa and chocolate, since flavanols are present in many other foodstuffs (Tuenter et al. 2018). Other compounds found in cocoa and chocolate, such as caffeine and methylxanthines, are known to affect cognition and alertness, although again, these chemicals are not restricted to cacao. One of the minor chemical constituents of cocoa and chocolate is an alkaloid called salsolinol. It is derived from dopamine and has been found to bind to the D_3-receptor, which plays a role in the reward system in animals (Tuenter et al. 2018). Some researchers have hypothesized that salsolinol's dopaminergic activity might be important in modulating the mood effects of cocoa and chocolate (Melzig et al. 2000). Salsolinol is found in cocoa and chocolate at concentrations of roughly 25 µg/g and it is reckoned that eating about 100 g (3.5 oz) of dark chocolate is enough to provide a pharmacologically relevant concentration (Tuenter et al. 2018). The only problem is that salsolinol is not very good at getting through the blood-brain barrier (Melis et al. 2015), which puts a bit of a downer on things. Still, you can cheer yourself up by savoring a piece of chocolate, since its orosensory properties are known to explain, in part at least, the desire to eat chocolate and the resulting mood-enhancing effects (Tuenter et al. 2018).

3

Chocolate Is a Product of the Cacao Tree

Anyone seeing a cacao tree for the first time would be forgiven for thinking what a strange tree it is. After all, its flowers, and subsequently its fruit—those large, round to oval-shaped pods—come straight out of its trunk. Botanically, this is known as cauliflory, a term translated literally as "stem-flower" and refers to flowers and inflorescences that develop directly from the trunks and branches of woody plants. Cauliflory is most prevalent in tropical rainforests, which happen to be the natural habitat of the cacao tree. Here, it grows in the shade of the lower story of the forest canopy. As I mention in the prologue, this is what I remember from my childhood: seeing, in the dappled light of a cacao plantation, small trees with large red or yellow pods sticking out of their lichen-encrusted trunks. Only later at university did I learn that in the lower reaches of the forest canopy, flowers and fruits that are borne on trunks and tree limbs are easily accessible by pollinators and frugivores (fruit eaters), including birds, bats, climbing mammals, and insects.

Given the name *Theobroma cacao* by Linnaeus in 1753, the cacao tree belongs to a group of small trees that can be found growing wild in the Amazon basin and tropical areas of South and Central America. The cacao tree is thought to have begun its evolutionary journey in the upper regions of the Orinoco and Amazon rivers, in South America. Although wild cacao can be found in Central America, this probably represents an extension of its geographical range from South America (Young 2007, 3). In the Amazon, there is

Figure 3.1. Pods on a cacao tree in Ghana. By permission of Harry Evans.

much variation in the shape, surface texture, and color of the pods, suggesting a considerable genetic diversity among wild populations of the cacao tree. This variability decreases as you move down the Amazon from its center of origin on the slopes of the Andes. There are more than 20 species within the genus *Theobroma*, but the cacao tree is the only one cultivated widely.

Although the genus *Theobroma* is many millions of years old, Young suggests that the cacao tree might have evolved more recently, perhaps just 10–15 thousand years ago, coinciding with the arrival of the first humans in South

America (Young 2007, 3). Although today we are interested in cacao beans, ancient peoples are likely to have selected cacao for its pulp. Within the formidable exterior of the cacao pod, its seeds (we call them beans) are surrounded by a white, sweet-tasting pulp. This nutritionally rich pulp is much sought after by monkeys, squirrels, rats, and bats, which devour it, discarding the bitter-tasting seeds. The delicious pulp is the reward to the animals for liberating the seeds from the pods, which, even when ripe, remain on the tree (Young 2007, 3).

Thousands of years ago, a natural population of *T. cacao* was spread throughout the central part of the Amazon region, moving west into Guiana and north to southern Mexico (Cuatrecasas 1964). In time, and separated by the Isthmus of Panama, this population of the cacao tree diverged into two distinct subspecies: Criollo in Central America and Forastero in South America (Young 2007, 15). The Spanish considered Criollo to be the original type of cacao, hence the name, which in Spanish means "of local origin" or "native." Forastero cacao was thought to be non-native, which is reflected in the Spanish name, meaning "foreigner" or "outsider." Criollo trees produce beans which are rounded and white in cross-section. Chocolate made from these beans is of very high quality: aromatic but lacking in bitterness. Unfortunately, Criollo trees are susceptible to a range of diseases, which has had a huge impact on its cultivation. So much so that Criollo cacao has become scarce and currently represents less than 3% of the world's cacao production. By contrast, the Forastero subspecies is widely grown and accounts for about 85% of world cacao production. Forastero varieties are hardy, vigorous, and more disease resistant than Criollo cacao. They produce beans that are smaller and flatter, with violet-colored cotyledons, and although beans from these types possess a stronger flavor with a higher fat content, the flavor is not generally considered of high quality. Having said that, there are exceptions, one of which is the Ecuadorian 'Cacao Nacional,' which produces fine-flavored cocoa of high quality (Wood and Lass 1985; Young 2007, 44–47).

The Start of Cacao Cultivation

Recent evidence indicates that the cacao tree was first domesticated around 5,300 years ago in southeast Ecuador, making this area the oldest site of cacao domestication yet identified (Zarillo et al. 2018). Domestication is one thing, but taking the next step to cultivating the cacao tree as a crop is quite another. Cultivation of the cacao tree is thought to have started about 2,000 years ago in Central America (Young 2007, 15–16). There is no evidence that it was cultivated in South America at this time. In fact, there is no mention of cacao cultivation in the cultural history of the Andean peoples of South America, and before the Spanish conquest, there were no cacao plantations in South America or the Caribbean. It is thought that the type of cacao grown in Central America must have belonged to the Criollo type, since there is no evidence that Forastero cacao was cultivated in that region (Wood and Lass 1985, 1). This might have been because Criollo beans can be made into a palatable drink with little or no preliminary fermentation, whereas beans of Forastero cacao need to be fermented for several days prior to use (Wood and Lass 1985, 3).

In Mesoamerica, cacao is cultivated in regions that are hot and humid with high annual rainfall and little in the way of a dry season. The tree thrives in deep, well-drained soils, with shade provided by larger trees, such as *Inga* and *Erythrina*. But there is evidence that the Mayans cultivated cacao in dry parts of Mexico such as the Yucatán. The trees were planted in *cenotes* or sinkholes, which were kept naturally wet and humid by circulating groundwater. Cacao trees can still be found in cenotes today and may be descended from cacao trees cultivated by the ancient Mayans (Young 2007, 23–26).

Prior to the Spanish conquest, native Indians had migrated from the central highlands of Mexico into Central America. Among the migrants were Pipil and Nicarao Indians, the former moving into what are now El Salvador, Honduras, and Guatemala, and the latter settling in Nicaragua. Cacao cultivation was important to these peoples and they developed irrigation systems that allowed them to grow the trees in areas with extended dry seasons and low rainfall. The newly arrived Spanish encountered cacao "orchards" distributed widely throughout Mexico and Central America, with extensive cultivation in the coastal region of the Gulf of Mexico, the southwest Pacific coast of Mexico, and parts of Guatemala, El Salvador, and Costa Rica (Young 2007, 26–29).

Following the Spanish conquest, cacao cultivation spread from Mexico to the Caribbean islands and parts of South America. Cultivation of cacao started in Venezuela in the sixteenth century, and although cacao was first planted in Trinidad by the Spanish in 1525 (Bekele 2004), a subsequent introduction was made from Venezuela in 1678, with seed of a Criollo type. At about this time, cacao was also introduced to Martinique and Haiti. In about 1600, Criollo cacao was taken by the Spanish across the Pacific to the Philippines, where 'Java Criollo' cacao was developed. Cacao cultivation spread later from the Philippines to Sulawesi and Java, and in 1798 Criollo cocoa was introduced to India and Sri Lanka from the Moluccas, off Indonesia (Wood and Lass 1985, 3–4).

Most cacao grown in the sixteenth and seventeenth centuries was of the Criollo type. Cultivation of Forastero cocoa started in the eighteenth century, and Brazil and Ecuador were the first countries to grow and use it. The fledgling cacao industry in Trinidad, which was based on the Criollo type, was almost completely destroyed in 1727 by what was then called a "blast." Although this might have been a hurricane, it is now considered more likely to have been a disease such as *Ceratocystis* wilt (see chapter 9), to which Criollo cacao is very susceptible (Bekele 2004; Wood and Lass 1985, 3–4). To resurrect the industry, Forastero cacao was introduced from Venezuela in 1757. Over time, this was interbred with the remaining Criollo cacao to produce a hybrid referred to as Trinitario (Young 2007, 23–26), widely considered to produce cocoa of elite quality (Yang et al. 2013).

Cacao planting in the Amazon basin started in the seventeenth century, but it wasn't until 1746 that the first plantings were made in the state of Bahia in Brazil, apparently by a French planter who brought cacao beans from the state of Para. These seeds are thought to have been derived from wild Amelonado cacao (a Forastero-type cacao from the Lower Amazon) in the Guianas and gave rise to the type of cacao known as Comum in Bahia. Cacao cultivation started in Ecuador in the seventeenth century and was of the Nacional type, which presumably arose from wild trees selected for seed from the early plantings (Wood and Lass 1985, 4).

There was considerable trade between Brazil and West Africa in the nineteenth century, and after Brazil gained independence from Portugal in 1822, Amelonado cacao was taken from Bahia to the small island of Príncipe off the West African coast. From there, it spread to the islands of São Tomé in 1830 and Fernando Po in 1854 (Wood and Lass 1985, 4; Young 2007, 40–44). Later in the nineteenth century, cacao was taken from there to Ghana and Nigeria, forming the basis of cacao cultivation in West Africa. Although Amelonado cacao became the dominant type grown in West Africa, Cameroon was an exception. In the mid-1880s, Cameroon was a German colony, and during this period, Trinitario cocoa was imported from South America and the West Indies. As a result, different Trinitario cultivars became established, especially in the east of the country, where Trinitario and Amelonado cacao were interbred. In 1945, cacao from Trinidad was sent to Ghana. The cultivars sent were based on material taken to Trinidad from the Amazon basin as part of a search for resistance to witches' broom disease. Cacao production in West

Africa since 1960 is based largely on these cultivars, with more recent additions of new material from the Cocoa Research Unit at the University of the West Indies in Trinidad. Cacao was also taken from Trinidad to Ceylon (now Sri Lanka), first in 1834 and 1835, and later in 1880, and from Ceylon it was taken, across the Indian Ocean, to Singapore, Fiji, Samoa, and Queensland in Australia (Young 2007, 40–44).

Today, world cacao production is dominated by West Africa, with just two countries, Côte d'Ivoire and Ghana, accounting for roughly half of all cacao produced. The largest producer by far is Côte d'Ivoire, which produced 1.581 millon tonnes in 2015/2016, with Ghana's production lagging behind at 778,000 tonnes. Ecuador and Brazil produced 141,000 and 232,000 tonnes (ICCO 2018), respectively, in the same period, while Trinidad, which produced more than 30,000 tonnes a century ago, has seen production plummet to just 500 tonnes in recent times (WIPO 2017).

Globally, cacao production in 2015/16 was 3.99 million tonnes (ICCO 2018). This represents an increase of 455,000 tonnes from 2008/09, when global production was 3.5 million tonnes. Apparently, this is equivalent to the weight of a line of double-decker buses stretching more than three times the length of Britain (Kew Science, n.d.). Now, that is a lot of chocolate.

Growing Cacao

Cacao is a tropical plant and, as such, the limits for its cultivation are 20°N and S of the equator, although most cacao is grown between 10°N and S, usually at elevations below 305 m. Having said that, it can be grown at higher elevations, for example, in the Venezuelan Andes, where it can be found at 1219 m (Purseglove 1968).

Growing wild in the understory of the tropical rainforest, the cacao tree is small and spindly, reaching a height of around 8 m. Here, it can be found in groups along river banks, where its roots may be under water for several months every year. In this environment of heavy rainfall, uniform temperature, high humidity, and dense shade, the tree flowers sparsely and carries little fruit. Although cacao can survive levels of shade that would kill other species, it can grow well with only light shade. Because the rate at which leaves can make sugars (known as photosynthesis) is reduced when light levels are low, plants can compensate by producing larger leaves. In dense shade, the cacao tree produces bigger leaves with a larger leaf area, allowing the plant to trap as much precious light as possible. Under cultivation, young cacao seedlings grow best if shading is used to reduce light levels to

Figure 3.2. Cacao trees with pods in Ecuador. By permission of Harry Evans.

around 25% of full sunlight. This level of shade is required not just to reduce light intensity, but also to buffer the microclimate, to ensure that the young seedlings don't suffer from water stress (Wood and Lass 1985, 171). When the seedlings have grown into young trees, light can be increased to about half of full sunlight, and according to Purseglove (Purseglove 1968), under optimal conditions of rainfall and nutrition, mature trees can tolerate full sun. Perhaps this should not be a surprise, because as the trees mature and their canopies develop, those of neighboring trees meet, decreasing the need for shade (Wood and Lass 1985, 171). In fact, yields of mature trees are usually higher when grown with little or no shade (Wood and Lass 1985, 8, 171). Maintaining these high yields, however, requires larger leaves and high rates of photosynthesis, which, in turn, requires good nutrition. In other words, cacao trees growing under light shade or no shade require fertilizer application (Wood and Lass 1985, 171).

Water is also important for the cacao tree, which is said to require between 1400 and 2000 mm/year of rainfall for normal growth (Wood and Lass 1985, 38–79). Anything less than 1200 mm/year is likely to lead to water stress (Alvim 1977). What's important for the cacao tree, however, is not the total amount of rainfall it receives, but the distribution of rainfall over the year. So, growth and yield suffer if cacao trees are subjected to three months where rainfall falls below 100 mm/month (Lahive et al. 2019). Juvenile cacao trees are particularly sensitive to water stress and fail to recover if subject to lack of water for too long. This sensitivity to water stress means that establishing young cacao plants in the field can be difficult following a severe dry season. Perhaps this should not surprise us, given that cacao originated in the Amazon basin, where water stress does not tend to be a problem.

Studies carried out more than 70 years ago concluded that the cacao tree does not like temperatures lower than 15 °C, with the absolute minimum being 10 °C (Erneholm 1948), although there is evidence that it can tolerate temperatures as low as 4–6 °C (Alvim 1977). In fact, temperatures in cacao-growing parts of the world lie between a maximum of 30–32 °C and a minimum of 18–21 °C (Wood and Lass 1985, 38–79). More recent temperature recordings reveal that maximum monthly temperatures in cacao-growing regions range from 26 to 34 °C, although daily maximums can be higher, as in Ghana, for example, where daily maximum temperatures of 35–44 °C have been recorded (Lahive et al. 2019). Temperature affects the rate of photosynthesis, in which the plant converts light and carbon dioxide into sugar. In cacao, the optimum temperature for photosynthesis has been recorded at between 31 and 35 °C (Lahive et al. 2019).

Flowers and Pollination Puzzles

Once the trees are two to three years old, they start to produce flowers, lots of them, on the trunk and along the branches. These are small, of a pentamerous design, and are borne on long stalks, or pedicels. The petals, which range in color from white to pink, have a curious shape: narrow at the base, expanding into a cup-shaped pouch, and ending in a broad tip, yellow in color and thought to provide pollinating insects with a landing stage (Young 2007, 92). There are 10 stamens, the male part of the flower, arranged in two whorls, an outer one consisting of five long, nonfertile staminodes and an inner whorl of five fertile stamens, each of which bears two pollen sacs (anthers) concealed within the cup-shaped pouch of the corresponding petal. The female part of the flower is located centrally. These make up the stigma, which receives the pollen grains, sitting atop a long stalk, known as the style. At the base of the style lies the ovary, split into five chambers, each containing many ovules around a central axis. The five red staminodes enclose the central style and stigma in a sort of circular fence, thought by Young to provide a barrier between the stigma and the anthers, making it difficult for pollen to be transferred from the anthers

Figure 3.3. Flowers of a cacao tree in Trinidad. By permission of Ashley Parasram, Trinidad and Tobago Fine Cocoa Company Ltd.

to the stigma of the same flower. Such an arrangement would promote cross-fertilization between different trees (Young 2007, 94–95).

Flowers open fully at dawn, having started to open, slowly, late in the previous afternoon. Shortly after opening, pollen is released from the anthers, and on this first day of flowering, the style and stigma are most receptive. This presents a very small window of opportunity for pollination, and flowers that are not pollinated, fall off the tree the next day. Very few flowers are successfully pollinated—fewer than 10%—and of those, only 2% grow into mature fruit (Young 2007, 95).

The cacao tree can produce as many as 125,000 flowers along its main branches (Toledo-Hernández et al. 2017). This is a prodigious number of flowers, and yet all this flower production results in very few pods. How can this be? One of the reasons lies in what is known as pollination self-incompatibility. Flowers on most plants are hermaphrodite, containing both female and male parts. With the anthers and stigma close together, self-pollination is a real possibility. Although common in many plant species, conferring advantages such as having a stable genotype and not having to rely on pollinators, there are disadvantages to self-pollination. One disadvantage is inbreeding depression, which reduces the vigor and biological fitness of the plant. Cross-pollination, on the other hand, introduces genetic variation into a population. Natural selection can act on this variation, weeding out individuals ill-adapted to changing environments, for example. It can also result in hybrid vigor. Of course, in the case of cacao, cross-pollination relies on pollinators, which means having to spend energy attracting them to the flowers, but to get the most from sexual reproduction, cross-pollination is the way to go. The problem is that in flowers with female and male parts in close proximity, this is difficult to achieve.

What to do? Well, it should come as no surprise that plants have evolved mechanisms to increase the chances of cross-fertilization, and in around 30% of higher plants, to prevent self-fertilization. The latter are known as incompatibility mechanisms. Self-incompatibility is a biochemical recognition and rejection process that prevents self-fertilization. This involves interactions between the pollen grain and either the stigma or the style. After a pollen grain lands on a stigma surface, it germinates, producing a germ tube. This germ tube needs to penetrate the stigma and grow down the style to reach the ovary. Incompatibility can occur on the surface of the stigma, before the pollen germ tube has penetrated it, or within the style, following penetration. Here, the pollen germ tubes do not develop properly and don't even reach the ovary. But in cacao, the mechanism is different: the pollen germ tubes develop normally, but the male and female gametes do not fuse. This is known as late-acting self-incompatibility (Wood and Lass 1985, 22–23; Ford and Wilkinson 2012).

The geneticist F. J. Pound, working at the Imperial College of Tropical Agriculture (ICTA) in Trinidad in the early 1930s, was the first to report self-incompatibility in cacao (Pound 1932). He demonstrated that certain trees could not set fruit with their own pollen nor with pollen from other trees of the same kind. In such an incompatible pollination, the ovary fails to develop, and the flower falls off the tree after three or four days. Since Pound's work, the existence of self-compatible and self-incompatible cacao trees has been established; however, the degree of incompatibility varies between different populations of cacao trees. For example, Amazonian cultivars are all self-incompatible, though they are generally cross-compatible, while the Amelonado population is entirely self-compatible. By contrast, in Trinitario cultivars, many trees are self-incompatible and will not cross with other self-incompatible trees. For successful pollination, these cultivars need pollen from self-compatible trees. According to Wood and Lass, in some Trinitario populations, cross-incompatibility may have limited yields, although this can be avoided with more modern hybrids by planting mixtures of different hybrids (Wood and Lass 1985, 22–23).

Cacao plantations typically grow mixtures of self-compatible and self-incompatible trees; therefore, many different varieties must be planted to ensure adequate cross-pollination for self-incompatible trees to set fruit. There might be less fruit-set in self-incompatible trees, but they tend to possess desirable agronomic qualities, such as improved disease resistance.

What Pollinates Cacao?

In cacao, successful pollination requires a minimum of 35–40 pollen grains to be deposited on the style, most likely by tiny midges belonging to the family Ceratopogonidae. Although cacao flowers are visited by many other insects, including ants, aphids, and fruit flies, their role in pollination is thought to be minimal (Toledo-Hernández et al. 2017; Wood and Lass 1985, 21–22).

The pollinating midges are more abundant in the rainy season than the dry season and are most active in the early morning and a few hours before sunset. During these periods, they forage flowers within a 5–6 m radius. Indeed, although they can fly for distances of up to 3 km, work in West Africa showed that most pollination takes place between neighboring trees (Wood and Lass 1985, 21–22). The midges, usually biting midges of the genus *Forcipomyia*, prefer cool, dark, moist habitats and breed in rotting vegetation, including cacao husks and dead tree trunks. On cacao plantations, this material is usually removed, thereby reducing the breeding habitat available to the midges. In fact, this removal is thought to be responsible for the low abundance of pollinating midges on cacao plantations. Working in Costa Rica in the early 1980s, Allen Young found that the abundance of several species of *Forcipomyia* midges increased greatly when disks of rotten banana stems were added to the ground litter. Intriguingly, although Young carried out experiments on two cacao farms, the effect was observed on one farm, but not the other. The increased midge numbers were recorded on a farm where shade cover was provided by an open canopy of bananas mixed with a variety of wild trees. Here, cacao trees produced plenty of flowers, and many midge species were recorded. On the other farm, a uniform cover of *Hevea* rubber trees provided the shade, cacao trees produced few flowers, and only a few midge species were found. The conclusion was that a cacao farm with heterogeneous shade cover would be expected to harbor a larger resident pool of pollinating midge species. Young was also of the opinion that midge abundance was a limiting factor in the pollination of cacao (Young 1982; Young 2007, 128–132).

Fast-forward 31 years and Young's findings relating to midge abundance and the presence of banana stem slices were confirmed by results from West Africa. Research in Ghana by Mike Adjaloo and colleagues demonstrated that providing additional, midge-specific substrates such as decomposing cacao leaf litter, slices of banana pseudostem and cacao pod husks led to increased fruit set (Adjaloo et al. 2013). Further support for the benefits of midge breeding sites was provided by work published in 2017 by Samantha Forbes and Tobin Northfield in Northern Australia, who found that addition of cacao pod husks increased the number of fruits per tree (Forbes and Northfield 2017b). It seems that by increasing the availability of suitable habitat for midges to breed, midge abundance can be increased, and with it, pollination success.

But midges aren't the only insects to pollinate cacao. Other invertebrate species represent almost half the visitors to cacao flowers, with ants and bees making up a large proportion of the non-midge visitors. Ants are thought to facilitate pollination indirectly by disturbing pollinators, making them move more frequently between flowers and enhancing pollination success. So, when Arno Wielgoss and colleagues excluded ants from cocoa trees in experiments in Sulawesi, Indonesia, fruit-set of cacao flowers was significantly reduced (Wiegloss et al. 2014).

Natural forests are important habitats for insects, including the midges that pollinate cacao. These insects can migrate from forests to nearby agroforestry systems such as cacao farms, and if areas of forest are removed, the diversity and abundance of insects decrease. When Young carried out his experiments in Costa Rica in the early 1980s, he observed that pollinators quickly colonized cacao trees near patches of forest. Perhaps forest conservation is necessary to ensure pollinator abundance.

Pod Growth

If pollination has been successful, a pod develops. It grows slowly for the first month or so, but then speeds up and reaches its full size after 4–5 months. Another month is then required for the pod to ripen, when its color may change.

Figure 3.4. Cacao pods. By permission of Ashley Parasram, Trinidad and Tobago Fine Cocoa Company Ltd.

Each pod will contain between 20 and 60 seeds (the cacao beans), with Forastero cultivars having more beans than Criollo cultivars.

Although only a small percentage of cacao flowers are successfully pollinated, trees usually set too many fruits to carry through to maturity. When faced with this problem, other tree crops jettison many of the fruits in a process known as fruit thinning. Cacao is no different in that it also operates "pod thinning," but here, instead of falling off the tree, the young pods (known as cherelles) remain on the tree. The cherelles stop growing and about a week later turn yellow, then black, and finally they shrivel. This process is known as cherelle wilt and usually results in the loss of about 80% of pods on a tree.

A cacao tree will start to produce significant numbers of pods by its fourth or fifth year and with proper care and management will produce 20–30 pods each year for 10 years, although trees can continue to yield for more than 30 years. Ripe pods can be found on trees at any time of the year, thanks to the continuous growing season in the tropics; however, most countries have two peak periods of production each year. Depending on the variety of cacao grown, each pod will contain between 20 to 50 beans. So, next time you start munching on your favorite chocolate, just remember that it can take all the pods produced by one tree to make a 450 g bar of chocolate (Cadbury, n.d.). While this fact sinks in, just think: 90% of the world's cacao is grown on smallholdings of less than five hectares, and yields vary across the globe (ICCO 2012). The average yield from a smallholder farm is 350 kg per hectare, but yields range from an average of 200 kg per hectare in Ecuador and 450 kg per hectare in Côte d'Ivoire to between 700 and 1500 kg per hectare in Sulawesi, Indonesia (Wessel and Quist-Wessel 2015). The low yields in some parts of the world (West Africa, for example) are due to the age of the trees, together with low soil fertility, and the often-devastating impact of diseases and pests (Wessel and Quist-Wessel 2015). In Indonesia, although cacao production has increased hugely in the last 20 years, average yields have remained at around 700 kg per hectare for some time (Agriculture Department of Indonesia 2008). This is due to a lack of improved varieties, as well as the effects of diseases and pests. Although cacao genotypes with resistance to some diseases have been identified by researchers, getting them through to farmers continues to pose a challenge.

The Cacao Tree Reclassified

Genetic improvement of the cacao tree is necessary if yields and resistance to diseases and pests are to be increased. To achieve these aims, plant breeders have traditionally crossed trees from the two main genetic groups, Criollo and Forastero. Hybrids of these two groups form a third genetic group, Trinitario, although some researchers argue that Criollo and Trinitario, like Nacional and Amelonado, are traditional cultivars and not genetic groups. What is needed is a thorough classification of *T. cacao* based on genetic data from different populations. When this was carried out using accessions of wild and cultivated cacao, it was estimated that up to 44% of trees in germplasm collections had been misidentified (Motilal and Butler 2003). This misidentification makes it difficult to work out population structure, and without this, breeding and management of the genetic resources of cacao are hampered (Motamayor et al. 2008).

Juan Motamayor and colleagues set out to put things right. To do this, they obtained 1,241 accessions covering a wide geographical sampling area in Latin America and determined the differences in genetic makeup of the individual accessions, in a process known as genotyping. Their results indicated the existence of 10 genetic clusters rather than two genetic groups (Criollo and Forastero), leading Motamayor and his colleagues to propose a new classification of cacao germplasm, which more accurately reflects the genetic diversity available to breeders (Motamayor et al. 2008). So, rather than the traditional classification as Criollo, Forastero, and Trinitario, the new genetic clusters are: Marañon, Curaray, Criollo, Iquitos, Nanay, Contamana, Amelonado, Purús, Nacional, and Guiana. New genetic clusters continue to be defined as our knowledge of cultivated and wild cacao increases (Zhang et al. 2012). This should help curators of cacao germplasm collections, geneticists, and plant breeders in their efforts to manage and exploit cacao's genetic resources more effectively (Motamayor et al. 2008).

Something else that should help breeders in their attempts to improve the cacao tree is the publication in 2011 of its draft genome. A consortium of researchers from across the globe sequenced and assembled the genome of *T. cacao*, which is housed on its 10 chromosomes; transposable elements, otherwise known as "jumping genes" because they move from one location on the genome to another, make up 24% of the cacao genome. One of the benefits of this work is the identification of a range of genes that will be useful in breeding for enhanced yield and effective disease and insect resistance (Argout et al. 2011).

. . .

When I last visited Trinidad in 1990, I took photographs as I walked through the cacao estate at Gran Couva. What is immediately obvious, apart from the green, yellow, and red pods sticking out of the trunks and larger branches of the trees, is the presence of flushes of red leaves among the otherwise green foliage. Newly formed leaves are red due to the presence of anthocyanin pigments, which decrease in concentration as chlorophyll develops and the leaves mature to a dark green color. Walking through the cacao then, in my early 30s, brought back vivid memories from my childhood—the shade and shadows, the peace and quiet, and the smell of moist, decaying leaves covering the ground beneath and between the trees. The Trinidadian novelist Sir Vidia Naipaul, who attended the same school as I, Queen's Royal College in Port-of-Spain, thought the cacao farm was like "the woods of fairy tales," while Allen Young's description (Young 2007, 90–91) is well worth quoting:

I, too, have experienced the mystique of old cacao plantings. There is something wonderful about a plantation of wizened old cacao trees. Within this setting, moisture drips from every leaf and branch, and the mulch smells steamy and fresh. Little pools of sunlight filter through the large, flat leaves of the cacao trees, illuminating the leaf litter with its earth tones of russet, yellow, orange, and brown. The dapple of sun and shade often traps for brief moments the to-and-fro passage of flying insects, many unknown and certainly unnamed by science, just above the leaf litter floor.

How brilliantly evocative is that?

I emerged from the shade of the cacao trees and their shade-giving companions to the sight of raised wooden platforms covered with drying cacao beans. This is a sight I remember well from my childhood. I was told that these drying beans, with their characteristic smell, would eventually end up in the chocolate I always craved. I found that difficult to imagine—from smelly beans to yummy chocolate? Surely not.

4

Transforming Cacao Beans into Chocolate Bars

When to harvest any crop is never as straightforward as it might seem to those of us who are not farmers. One of the factors likely to affect timing of the harvest is weather. In cacao, although pods take between five and six months to reach full ripeness, this can vary considerably depending on temperature. Basically, pods grow more slowly in the cooler months of the year, increasing the time to harvest. Rainfall is also important. We've already seen that most countries have two peak production periods during the year, although cacao tends to be harvested several times a year, and even weekly on large farms in some countries. In those countries with defined wet and dry seasons, the main pod harvest is usually five or six months after the start of the rainy season. Some countries, Malaysia for example, have no real dry season, and here peaks in pod production are less pronounced (Wood and Lass 1985, 446).

So, when are pods ready to harvest? Pods change color as they ripen: green pods turn orange-yellow, and red pods become orange, although when several varieties of cacao are being grown, identifying ripe pods by color can be difficult. It's just as well that pods can be left on the tree for up to a month after ripening before they need to be harvested. One of the problems with this is that the longer pods are left on the tree, the greater the chance of developing a rot. This can be a problem on small farms, where pods are harvested just three or four times during the harvest period; on larger farms, pods tend to be harvested weekly, thereby reducing the likelihood of pod rot.

When it comes to getting the pods off the tree, there are no shortcuts—harvesting cacao is slow and laborious. Pods are cut from the tree, using a cutlass or machete if the pods are within easy reach, or a special knife at the end of a long pole for pods higher up in the tree. Removing pods must be done carefully, to avoid damaging the flower cushions and providing an easy entry point for pathogens. Once they are off the tree, the pods must be opened to get at the beans enveloped in a mucilaginous blanket of gel. Straightforward enough, I hear you say. Well, not really. Do you split open the pods as soon as they are cut from the tree? Do you take the harvested pods to a central spot in the field and wait until sufficient have been collected before they are cut open? Or do you, perhaps, transport them out of the field to a covered area and chop them open there? The answer to this depends on where the cacao is being harvested. In West Africa and Trinidad, for example, pods are collected in the field over several days and opened when sufficient pods have been harvested. One problem with this approach is that any delay in opening the pods increases the risks of damage by pod-rotting fungi, although delaying pod opening for several days after harvest seems to speed up the fermentation process. Opening the pods as soon as they are cut from the tree makes it difficult to monitor for pod ripeness and quality, but it offers some advantages: since the discarded husks are distributed throughout the field, valuable nutrients are released back into the soil, and the decaying husks can provide breeding sites for pollinating midges (Wood and Lass 1985, 446–449; Toledo-Hernández et al. 2017).

There are several ways to open the harvested cacao pods. One commonly used method is to split them open using a machete or cutlass. I have seen this done in Trinidad by workers holding the pod in one hand and splitting it open with a cutlass. Clearly you need to be well-practiced at this. Apart from the health and safety implications, this process risks damaging the precious beans. Another approach is to break open the pods using a wooden club. The trick here is to strike the central part of the pod, whence it splits into two halves. As you might expect, machinery has been developed for opening cacao pods, but most smallholders still opt to open them by hand. Once the pod has been opened, the wet beans in their all-embracing pulp are scooped out, ready for the fermentation process. Apparently, an experienced cacao farmer can harvest and cut open 900 pods per day, and if the tasks of harvesting and opening are performed separately by two people, each can deal with 1,500 pods per day (Wood and Lass 1985, 446–449). That's an impressive work rate, especially if you are opening pods using a machete.

Figure 4.1. Cacao beans fermenting in Ghana. By permission of Harry Evans.

Figure 4.2. Tool for harvesting cacao pods. By permission of Ashley Parasram, Trinidad and Tobago Fine Cocoa Company Ltd.

Fermenting the Beans

What happens to the cacao beans after harvest has a big impact on the taste of the final product. Two steps are essential in developing the wonderful flavor of chocolate: the first is undertaken by the grower, who cures the beans, and the second is the province of the manufacturer, who roasts the beans.

Curing the beans is itself a two-stage process: fermentation followed by drying. The harvested beans are either packed into boxes, known as "sweatboxes," or heaped into piles and covered with mats or banana leaves and left for up to seven days. Beans removed from the pods are encased in a mucilaginous pulp, which is acid due to the presence of citric acid; it's also rich in sugars and, at least initially, sterile. The high sugar content and acidity of the pulp provide ideal conditions for a variety of microbes, including yeasts, which anaerobically convert the sugars into alcohol. Soon after fermentation begins, the cells in the pulp start to break down and this liquefied pulp, known as "sweatings," runs off. Some of the acidity is lost and the temperature within the fermenting beans increases, favoring the growth of bacteria, some of which convert the alcohol into acetic acid. The combination of heat, alcohol, and acetic acid kills the embryo within the cacao beans. When alive, the cotyledons within the bean contain small highly colored cells packed full of polyphenols. When the beans and the cotyledons within them die, the colored cells are ruptured, releasing the polyphenols, bringing them into contact with various enzymes. This converts the polyphenols into the precursors of the much sought-after chocolate flavor, while at the same time reducing the astringency and bitterness of the beans (Wood and Lass 1985, 451–458).

One of the polyphenols in cacao beans is tannin, which composes up to 15% of the weight of each bean. Tannins impart an astringent flavor to the beans and need to be removed so as not to detract from the flavor of the final chocolate product. Fermentation helps to remove tannins from the beans. Inadequate fermentation leaves too much tannin, requiring the chocolate manufacturer to remove the offending polyphenol, a process which can impair the flavor of the chocolate (Young 2007, 81–85; Aprotosoaie et al. 2015).

So, how long should cacao beans be fermented? That depends on the variety grown, because beans of different cacao varieties don't all ferment at the same rate. Forastero beans, for example, take longer to ferment—usually three to seven days—than beans of Criollo cacao, which take between two and three days (Wood and Lass 1985, 451–458; Young 2007, 81–85). This seems to be due to the naturally milder flavor of Criollo beans—it takes a longer fermentation to reduce the harsh flavor of Forastero beans.

There is no sidestepping fermentation if you want to ensure good flavor development of your cacao beans. Sure, you can roast raw or unfermented cacao beans, but they do not produce the characteristic cocoa aroma. This aroma, and the flavor, of cocoa and chocolate is due to a variety of volatile compounds of which more than 600 are present in fermented cacao beans (Aprotosoaie et al. 2015). Although many are found in the beans prior to fermentation, some are formed during the fermentation process. Researchers at the University of New South Wales in Australia found 40 volatile compounds in fermented beans that were not detected prior to fermentation (Ho et al. 2015). As well as providing food for microbes during fermentation, the sticky pulp surrounding the beans also seems to play a role in developing the flavor of the beans. In work published in 2018, Irene Chetschik and colleagues found that various aroma compounds migrate from the pulp into the beans during fermentation, with the pulp acting as a reservoir of particular aroma compounds as fermentation proceeds (Chetschik et al. 2018).

The smell of fermenting cacao beans is powerful. As a boy in Trinidad, I recall walking into a low-slung building containing sweatboxes filled with cacao beans. The smell was overwhelming. Some 25 years later, on a visit to a cacao farm in Gran Couva, the smell of fermenting cacao beans transported me straight back to my 8-year-old self. The smell was a strange mix of a brewery and a winery, with some vinegar thrown in. In his book *A Way in the World*, Sir Vidia Naipaul describes it as "an acrid smell of fermenting fruit . . . like the smell of maturing casks or vats of wine" (Naipaul 2011). It's difficult to reconcile the smell of sweating cacao beans with the delicious aroma of chocolate, but luckily for us, the ethanol and acetic acid by-products of fermentation do not make it through to the final product.

Drying the Beans

After fermentation, the beans must be dried. Drying reduces the moisture content to about 7%, making them less likely to go moldy during storage and subsequent transport to the chocolate manufacturer. There is, however, more to drying than preventing mold. It helps to further reduce the bitterness and astringency of the beans, as well as developing the chocolate brown color of properly fermented beans. But care must be taken with drying—get it wrong and the flavor and quality of the chocolate will suffer. There is a balance to be struck: if the drying process is too slow, molds might develop, penetrate the seed coat, giving rise to unwanted or "off-flavors," while rapid drying could prevent completion of the oxidative changes that began during fermentation, leading to excessive acidity (Wood and Lass 1985, 478–492).

Beans can be dried in the sun or artificially. In some countries in the West Indies and South America, beans are dried on wooden floors with movable roofs. The beans are spread out on the floor and turned frequently to ensure uniform drying. In sunny weather with little rain, beans can dry within a week, but if it starts to rain and the roof is moved over the beans to keep out the rain, drying will take longer. In West Africa, beans are dried on mats raised off the ground or on concrete floors. Where rain is likely following harvest, as in Brazil and southeast Asia, artificial drying might be necessary. If so, simple convection driers could be used, with beans placed on a permeable drying platform above a simple flue. Alternatively, beans could be dried using platform driers with heat exchangers. Whatever method is used, artificially dried beans can be of poor quality, either due to contamination from smoke or because the drying process has been too quick (Wood and Lass 1985, 478–492; Gutiérrez 2017).

One of my childhood memories is of people walking all over cacao beans as they dried on a wooden platform. I learned that this was "dancing the cocoa" and was performed to improve the appearance of the beans. And sure enough, the result was beans with a polished patina. To "dance" the beans, they were first sprinkled with water and then walked on for about 30 minutes. This was carried out in the early morning, leaving the rest of the day for the beans to dry.

With the drying complete, the beans are ready for the journey to the processor. Traditionally, cacao beans are shipped in jute bags containing 60–65 kg of beans, although bulk shipment of beans has increased in popularity, with beans loaded directly into a ship's cargo hold or in shipping containers containing a flexi-bag. The latter method is used by large processors who buy huge quantities of beans (Gutiérrez 2017).

The long journey from pods on a tree to chocolate is nearing its end. All that remains is to convert the beans into chocolate.

Getting a Proper Roasting

I've just made the process of making chocolate sound easy—you just turn the beans into chocolate. If only! Chocolate manufacture involves a series of processes aimed at developing flavor and achieving the right texture. The starting point is not at all complicated but is essential—cleaning. Before the beans can be roasted, any foreign material must be removed—after all, nobody wants to

Figure 4.3. Drying cacao beans in Ghana. By permission of Harry Evans.

bite into a chocolate bar only to find a stone, or a bit of string, or a piece of metal.

Technically, the next step in making chocolate is roasting the beans, but not necessarily. At some point, the beans must be winnowed to remove the outer shell of the bean, leaving the cotyledons, known as the nib. You can opt for one of three approaches: roast the beans and then winnow; winnow the beans and then roast; or winnow the beans, grind them, and then roast (Gutiérrez 2017).

Roasting is necessary to continue the flavor development that began with fermentation and drying. Beans or nibs are roasted in special ovens at temperatures of 105–120 °C for up to 70 minutes, the roasting time depending on whether the end use is chocolate or cocoa. As they roast, the beans or nibs take on a rich brown color and acquire the characteristic chocolate aroma and flavor. If the nibs were not ground prior to roasting, they are now ground by passing them first through a pin or impact mill, which transforms the nibs into cocoa liquor and then through a ball mill, which leaves the liquor soft and silky. The result is a thick chocolate-colored liquid known as "mass"—cocoa butter with a fat content of 53%–58%. At this point the processes required to make cocoa powder and chocolate diverge. For cocoa powder, the fat content is reduced using hydraulic presses. Decreasing the fat content to 22%–23% yields high-fat powders for use in making drinks, while reducing the fat content to 10%–13% gives low-fat powders used for flavoring cakes, biscuits, and ice creams. Many cocoa powders are made from cocoa mass or nibs treated with an alkali such as potassium carbonate. This process, known as Dutching or alkalization, was developed by Van Houten, and its purpose was to increase the dispersion properties of the cocoa powder when mixed with milk. It is also used to intensify color, but the process will also affect the aroma and taste of the cocoa (Gutiérrez 2017). Alkalization reduces the acidity and astringency of the cacao beans and cocoa mass and can help to improve and intensify the aromatic characteristics associated with cocoa. The process can, however, reduce the polyphenol content of cocoa powder by as much as 80% (Wong 2017). Although cocoa powder tends to be made using alkalized cocoa mass, chocolate is usually made with nonalkalized cocoa mass (Gutiérrez 2017).

To make plain chocolate, the thick mass is mixed with sugar and cocoa butter, with just enough of the latter to allow the chocolate to be moulded. To ensure that all the solid particles within the chocolate are coated with fat, the mixture is subjected to a process known as "conching," named after the tank in which the process was originally carried out, which was shaped like a shell (Gutiérrez 2017; Wood and Lass 1985, 593). The chocolate is stirred continually at a temperature of 30 °C for several hours, which reduces acidity further by driving off volatile acids, makes the chocolate homogeneous, and finishes the development of flavor.

Milk chocolate for consumption in the United States and much of Europe is made using the same processes as for plain chocolate, but with the addition of milk powder. For the United Kingdom market, a different method is used: fresh milk is condensed with sugar, mass is added, and the resulting mixture dried under vacuum. This generates a product known as "crumb," to which extra cocoa butter is added to yield milk chocolate. Prior to use, both plain and milk chocolate must be tempered—getting the chocolate to the correct working temperature while ensuring that the crystalline structure of the cocoa butter in it is stable. This imparts a satin gloss to the chocolate and gives it its delightful melt-in-the-mouth properties. Fail to get tempering right and the chocolate will be gray and unappetizing (Gutiérrez 2017).

What Makes Chocolate Taste So Good?

The unique flavor of chocolate is determined by the variety of cacao, but it takes fermentation to release and develop the flavor potential locked within the beans, with the final touches provided by roasting and subsequent processing. Cacao beans contain a lot of fat, up to 32%, as well as sugars, proteins, the alkaloids theobromine (~3%) and caffeine (~0.2%), and polyphenols, which can compose up to 15% of the dry weight of the bean (Aprotosoaie et al. 2015; Ozturk and Young 2017). Just how much of these chemical constituents is present in the beans depends not just on the cocoa variety, but also on the conditions under which it was grown. The sugars in the gel-like pulp are converted into ethanol as well as lactic and acetic acids during fermentation, with further chemical reactions giving rise to the precursors of the chocolate flavor. Theobromine, caffeine, and polyphenols contribute to the bitter taste and astringency of roasted cocoa, while fats and volatile aromatic terpenes are important contributors to the flavor of cocoa.

Volatile compounds are major contributors to the flavor of cocoa and chocolate and, as we saw earlier, some 600 have been identified as important (Aprotosoaie et al. 2015). Among them are alcohols, aldehydes, esters, and pyrazines. The last group are key components of cocoa aroma, displaying nutty, earthy, roasty, and green aromas, and 80 of them are known to contribute to the overall flavor of cocoa (Magagna et al. 2017). Levels of pyrazines are higher in well-fermented cocoas from Ghana than Mexican cocoas, and Criollo cocoas have more pyrazines than Nacional cocoa. Esters confer a fruity flavor, and one ester, 2-phenylethylacetate, is mainly responsible for the unmistakable aroma of cocoa mass from Asia. Alcohols confer a fruity, green, floral aroma, and a high alcohol content is important if you want cocoa products with flowery and candy notes. Linalool is a major alcohol in roasted cocoa nibs, and it confers a flowery, leafy, and tea-like aroma. It is present in greater amounts in flavor-grade cocoas from Trinidad, Ecuador, and Venezuela, than in basic grade cocoa from West Africa or Malaysia. Aldehydes and ketones are also important for the development of good flavor in cocoa. They are usually formed during roasting, although small amounts of aldehydes may be formed during fermentation. Aldehydes such as 2-methylbutanal and 3-methylbutanal produce malty and buttery notes in both unroasted and roasted cocoa, while the ketone acetophenone produces sweet, floral aromas (Magagna et al. 2017).

Finding out what exactly is responsible for the characteristic aroma and flavor of chocolate has occupied scientists for decades. Although we now know that 600 volatile compounds play a role in the aroma of chocolate, are all 600 necessary for people to identify the smell as that of chocolate? According to Peter Schieberle at the Technical University in Munich, the answer is no. He and his colleagues found that people can recognise the chocolate aroma if just 24 of the 600 volatile compounds are present. These compounds included some imparting caramel-like, honey-like, and malty notes, as well as others responsible for the smell of cooked cabbage, for example. Individually, and in various combinations, these different volatiles did not smell of chocolate, but when all 24 were present, people were reliably fooled into thinking that they had smelled chocolate. Schieberle refers to these 24 volatiles as chocolate's chemical signature. This knowledge will help researchers like Schieberle to fine-tune fermentation and roasting to develop even more yummy-tasting chocolate (Arnold 2011; Frauendorfer and Schieberle 2006). I can hardly wait.

The flavor developed from cacao beans is influenced by a range of factors, including cacao variety and the fermentation process. But as we've seen above, off-flavors can arise under certain conditions, for example, as a result of under-fermentation and acidity. Mold can also produce off-flavors, and internal mold is particularly important because it cannot be removed during the manufacturing process. Within affected cacao beans, molds increase the content of free fatty acids, which can affect the flavor of the chocolate (Wood and Lass 1985, 507). Pod-attacking microbes such as the black pod pathogen, which we will deal with in the next chapter, can also attack beans, increasing free fatty acid content considerably, with consequences for chocolate flavor (MacLean 1953).

5

Small but Deadly

For almost as long as there has been life on this planet, there have been parasites—organisms that live off the hard work of others. Plants are used to being at the bottom of the food chain—performing the miraculous feat of converting carbon dioxide into sugars using sunlight as fuel—and along come the biological ne'er-do-wells and help themselves. But being a parasite isn't easy and plants are no push-overs. Plants are bristling with defenses (Walters 2017), and would-be parasites need to deal with these if they are to get to the nourishment within plant cells. Over time, successful parasites evolve the means to avoid detection, thereby evading plant defenses, and with the passage of yet more time, plants upgrade their ability to detect parasites, resulting in an unending game of evolutionary tit for tat. Plants and parasites coevolve, developing complex and intimate relationships in the process.

The cacao tree is no exception, and in its native range in the headwaters of the Amazon, several parasites coevolved with cacao, including the organism responsible for witches' broom disease. Frosty pod rot, which also affects cacao, coevolved with a relative, *Theobroma gileri,* in the Choco forest on the Pacific coast of Colombia and Ecuador (Harry Evans, email to author January 2019). At this point in their evolutionary history, the parasites causing these diseases have the upper hand and inflict huge losses on the cacao crop in South America. When cacao was taken across the globe to West Africa and Asia, it left these parasites behind, and in the new environment, and free of its old enemies, cacao thrived. But it didn't take long for parasites to move in on the new kid on the block, including the cacao swollen shoot virus in West Africa and the agent responsible for vascular streak dieback in Asia. Black pod, the disease that first captured my imagination as a boy in Trinidad, is widespread across cacao-growing regions of the world but is caused by different agents in different parts of the world.

Although cacao is afflicted by many diseases, these five are considered the most serious and together are responsible for considerable crop losses. Estimating these losses is difficult, but conservative estimates put the figure at 20% (Ploetz 2016). Given that total world production in 2012 was just over 5 million tonnes, this represents a loss of around 1 million tonnes due to diseases alone. In fact, diseases are the most serious biological constraint to cacao production, but matters would be much worse if diseases currently restricted to certain parts of the cacao-growing world were to spread. It's time we looked at these diseases more closely, and let's start with black pod, a disease estimated to cause annual losses to the global cacao crop of some US$4 billion (Guest 2007).

Black Pod

Cacao pods afflicted by this disease *do* turn black, as the name implies, but the initial symptoms are less conspicuous. Careful inspection of a pod just a couple of days after the parasite has managed to evade the pod's defenses reveals a small translucent spot, marking the early stages of the progress of this devastating disease. Things move quickly, for the small spot soon turns brown, expanding as it does so, before turning black. With the edge of the brown/black lesion advancing at the rate of 12 mm a day, the whole pod can turn black within two weeks (Wood and Lass 1985, 270–271; Surujdeo-Maharaj et al. 2016, 233). The changes are not restricted to the surface: within the pod, its tissues, including the precious beans, shrivel, leaving a black, mummified pod hanging from the tree. The rapid progress of the disease can render the pods commercially useless within three weeks.

Although attack on cacao pods is the most common sign of the disease, no part of the tree is safe, since the pathogen also infects stems and trunks, flower cushions, roots, and even leaves. A flower cushion is the term used to describe a leaf axil which has produced flowers and pods for several years and become thickened as a result. Mycelium of the pathogen can spread from infected pods, along the stalk, into the flower cushion and further along the stem, giving rise to a canker. If the pathogen has entered the tree via a wound, this can also give rise to a canker. Irrespective of its origin, a stem canker can disrupt the function of the vascular tissues below the bark, which includes transport of carbohydrates in the phloem and water in the xylem, leading to defoliation and death of twigs and even the whole tree. Infection of leaves leads to browning of tissue as cells are killed, disrupting photosynthesis and reducing the production of carbohydrates. This is bad news for the tree, since fewer carbohydrates means less energy and fewer building blocks to support growth of new leaves and pods. Root infection has received little attention, but it does occur. This should come as no surprise, since several species of *Phytophthora* are highly aggressive destroyers of roots in many plants.

The agent responsible for black pod is full of surprises. For a start, it is related to the organism that causes potato late blight, the destroyer of potato crops across Europe in the mid-nineteenth century and the cause of the Irish potato famine of 1845–1846. Blight had a devastating impact on the Irish population, leaving at least 1 million people dead and leading to the emigration of some 1.5 million more, mostly to the United States and Canada (Schumann and D'Arcy 2012). Potato blight struck again in 1916, with nearly three-quarters of a million German civilians dying: potato crops could not be protected because the copper required to treat the crops was needed to make bullets in World War I. More than 100 years later and potato blight is still a serious threat to potato crops, thanks to the speed with which the pathogen evolves.

In the middle of the nineteenth century, plant diseases were thought to be caused by "bad air," and the idea that they might be caused by a fungus was considered heretical (Griffith 2007). As potato blight started to ravage crops in the summer of 1845, a French cleric, Edouard van den Hecke, observed a fungus growing on blighted leaves and gave it the name *Botrytis*. Although there was support for the idea of a fungal cause for potato blight, for example from Camille Montagne in France, who named it *Botrytis infestans*, there was considerable opposition to the idea by the scientific establishment in Europe. Opposition continued until 1861, when the German botanist Anton de Bary described the life cycle of the pathogen, eventually giving it the name *Phytophthora infestans* in 1876. The name certainly fits the bill: it is derived from the Greek *phyto*, meaning plant, and *phthora*, meaning destroyer (Surujdeo-Maharaj et al. 2016, 214).

But with *Phytophthoras*, all is not as it seems, for these organisms are not

Figure 5.1. Cacao pods on a tree in Ghana showing symptoms of black pod disease. By permission of Harry Evans.

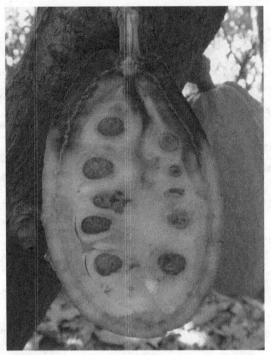

Figure 5.2. Canker caused by black pod infection of a cacao flower cushion. By permission of Harry Evans.

Figure 5.3. Canker on the trunk of a cacao tree caused by black pod infection. By permission of Harry Evans.

fungi at all. They might look like fungi, but they are more closely related to diatoms and brown algae. They are, in fact, more like plants in disguise as fungi and are an example of convergent evolution, where organisms from different lineages share similar features. *Phytophthoras* belong to a group of organisms called Oomycetes, also known as "water molds," and includes some of the world's most devastating plant pathogens. The group also includes pathogens belonging to the genus *Saprolegnia,* which causes diseases of aquatic organisms such as fish and crayfish. One member of this genus, *S. ferax,* is linked to the dramatic decline in amphibian populations across the globe (Heffer Link et al. 2002).

When I was an undergraduate in the mid-1970s, black pod was thought to be caused by a single species of *Phytophthora, P. palmivora.* Even then, however, it was already clear that the situation was far from straightforward. I had chosen black pod as the special study in my final year at Wye College and had been in touch with Mike Griffin and Philip Gregory regarding this scourge of cacao. Both were very helpful, sending me papers and, in the case of Gregory, even books, and they took the time to respond in detail to my queries. Mike Griffin, then working for the Agricultural Development and Advisory Service in Leeds, had been part of the Black Pod Research Project, based at the Cocoa Research Institute in Ibadan, Nigeria. He realised that I "had not clearly grasped the complex situation which exists with respect to *P. palmivora* and cacao" and proceeded to put things right. He informed me that there were at least three and possibly more, distinct *Phytophthora* species within the *P. palmivora* complex. At that time (1977), these were known as MFs (morphological forms), and Griffin was adamant that three of them were not different strains, but different species, which had yet to be named. This was important, because, as he pointed out in his letter to me, one of the main factors affecting the epidemiology of black pod disease on cocoa is the type of MF involved.

Mike Griffin was right. Over the subsequent decades, it became clear that each continent has a complex of species able to induce black pod in cacao. The main pathogen in West Africa turned out to be *P. megakarya,* while in the Americas and the Caribbean, *P. capsici* is important. Another species, *P. citrophthora,* is economically important on cacao in Brazil, and also affects cacao in Indonesia and India. Only *P. palmivora* is found in every country where cacao is grown (Brasier and Griffin 1979). Thankfully, not all species are present in all cacao-growing regions, but this could change. If it does, it would be catastrophic for the industry.

The Life and Times of the Black Pod Pathogen

As you might expect from something that disguises itself as a fungus, the black pod pathogen produces fungus-like structures. So, it grows like a fungus, producing thin threads, known as hyphae, that branch and spread in the tissues of the cocoa pod, forming a mycelium. Under the right conditions, which include high humidity and a nice, cozy temperature of 25–30 °C, the pathogen will start to reproduce. Mostly, this does not involve sex, and results in the formation of microscopic structures called sporangia, which are borne on stalks and appear as a fine, powdery white covering on the pod surface. Rather like Swiss army knives, these sporangia are multifunctional: they can produce a new mycelium, more sporangia, or they can produce zoospores. Formation of the spores is a rapid affair, taking just a few minutes—it's so quick it is thought to be one of the fastest developmental processes in biology (Walker and van West 2007). Providing the pod surface is wet, the sporangia open, liberating the zoospores. Courtesy of two flagella, one of which looks like a piece of miniature Christmas tree tinsel, these tiny spores swim around the pod surface, in search of a suitable entry point. Although a zoospore can swim for hours, it usually stops after about 20 minutes, transforms itself into a cyst and ditches its flagella, whereupon it germinates, producing a thin hypha which it uses to get through the pod surface. Now, there are two options available to the zoospore at this point: find a naturally occurring opening such as a stoma, or push straight through the pod surface. Stomata (plural of stoma) are usually located on leaves, where they open during the day to allow carbon dioxide to enter for photosynthesis and close at night to minimize water loss. But stomata are also found on other plant organs that are green and can photosynthesize, including cocoa pods. Many plant pathogens enter their host plant via a stomatal opening, including the black pod pathogens *P. palmivora* and *P. megakarya*. The other option—pushing directly through the pod surface—is also used by *P. megakarya*, but less so by *P. palmivora* (Ali et al. 2016). In this approach, the pathogen first produces a structure to help it through the pod surface—an appressorium. This attaches tightly to the plant surface and uses pressure to force a thin hyphal thread through into the plant. Sometimes the point at which the infection hypha is to penetrate is first softened up using enzymes, making it a little bit easier for the infection hypha to push through.

Some plant pathogens have a very subtle, intimate relationship with their host. Once they've entered the plant, they make their way gently into the plant cells, pushing through the cell wall and indenting the plasma membrane, but never piercing it. The result is a structure, called a haustorium, which the pathogen uses to obtain food from the plant's cells without killing it in the process. This is known as biotrophy and contrasts with necrotrophy, in which the pathogen does away with the softly, softly approach and simply blasts its way into the plant, killing its cells as quickly as possible, and lapping up the liberated nutrients. Some pathogens lie in between these two approaches and are known as hemibiotrophs. The black pod pathogens fall into this category—they are microbial con artists: they start off all nice and gentle, but quickly get fed up and kill the cells to get at the food. Witness the speed with which the disease spreads in the cacao pod—from healthy to useless in under three weeks.

It doesn't take long before the pathogen, having feasted on the nutrients within the dead and dying tissues of the cacao pod, emerges at the pod surface and produces more sporangia. The black pod pathogen likes rain, and for good reason, for the newly produced sporangia can be carried to other pods by rain splash (Maddison and Ward 1981). Two-thirds of new pod infections start in this way, but the pathogen can spread in other ways too. Husks of cacao pods falling to the ground add the microbe to the soil, where both *P. palmivora* and *P. megakarya* can survive for many months. In West Africa, where *P. megakarya* is the main cause of black pod, the pathogen can infect roots of the cacao tree, producing sporangia, which then liberate zoospores into the soil. These tiny bandits swim toward the soil surface, where rain can splash them up and onto the cacao tree. Once they land on a cacao pod, the cycle starts again and the spread of the pathogen through the cacao crop continues (Surujdeo-Maharaj et al. 2016, 237).

Among the material I received from Philip Gregory in late 1977 was a copy of the Cocoa Growers' Bulletin from 1973. In it was an article by Harry Evans, who was based at the Cocoa Research Institute in Tafo, Ghana, on a four-year secondment from the UK Ministry of Overseas Development (Evans 1973a). The article fascinated me. Harry described his work looking at the epidemiology of the black pod pathogen, *P. palmivora,* and suggested that a cocoa farm could be visualized in two dimensions. The first dimension concerns the individual cacao tree, with pods at different levels, which poses a problem for vertical dispersal of the pathogen within that tree. The second dimension relates to the farm and horizontal spread of the pathogen from tree to tree. Although vertical dispersal of the black pod pathogen within a tree could be explained by rain splash, accounting for the horizontal movement of the pathogen from

Figure 5.4. The West African Cocoa Research Institute (WACRI) at Tafo in Ghana. Initially established in 1938 as the Central Cocoa Research Station at Tafo, it became WACRI in 1944, with a substation at Ibadan in Nigeria. It is now the Cocoa Research Institute of Ghana (CRIG). By permission of Harry Evans.

tree to tree within a farm, and even to other farms, was more difficult to explain. If rain splash was responsible, you might expect what plant pathologists call an infection gradient: that infection levels would be high near the source of the pathogen—a tree with infected pods—but would decrease the further you moved from the infected tree. Researchers could find no evidence for an infection gradient within cacao farms, nor could they find sporangia of the pathogen in air samples taken on the farms. So, how was the black pod pathogen getting to distant cacao trees, either within a farm or on nearby farms? Might invertebrates be involved?

Harry was not the first person to look at invertebrates and the spread of the black pod pathogen. This accolade goes to Major Harry A. Dade who, having survived the horrors of the Somme in 1916, where he was badly wounded, went on to become a mycologist in the Colonial Service. He was posted to Ghana (then called the Gold Coast) in 1921 and over the next 15 years, published widely on diseases of cacao. Dade collected insects from cacao trees infected with black pod and allowed them to walk over sterile agar in tubes. When these tubes were examined later, the pathogen was found growing over the agar (Dade 1927). Dade concluded that "insects are probably the chief agents in the transportation of spores from one tree to another." This was exciting stuff, but as with everything in science, confirmation was required. Unfortunately, it took more than 40 years before this topic was revisited. Toward the end of the 1960s, the American plant pathologist August Gorenz was based at the Cocoa Research Institute in Nigeria. He was aware that ants that tended scale insects on cocoa pods were implicated in taking the black pod pathogen from the soil to pods, where they used the soil to build their tents. He tested this by placing diseased and sporulating pods at the base of two trees that had pods with scale insects being cared for by ants. Two weeks later, the pods all had symptoms of black pod infection, starting at the petiole, where the ants had built their tents. Three adjacent trees with pods at the same height, but without ants, showed no signs of black pod (Gorenz 1969). Subsequent work by Gorenz's colleague in Nigeria, E. K. Okaisabor, demonstrated that various ants could carry spores of the black pod pathogen passively on their bodies, transporting them from infected to uninfected pods (Okaisabor 1971).

At this point, Harry Evans enters the story. In meticulous research conducted in the early 1970s, Harry demonstrated that cacao trees colonized by tent-building ant species had significantly higher levels of black pod than ant-free trees. One species of ant, *Crematogaster striatula,* was thought to be particularly important because of its habit of using cacao pod tissue infected with black pod to construct tents. The ants didn't restrict their visits to infected pods—they were also keen on visiting wounded pods. This is important, since spores carried by ants can gain easy access to pod tissue via wounds. Harry's experiments showed that this ant can carry spores of the black pod pathogen not just within a tree but also between trees and could easily be responsible for the long-distance movement of the pathogen, even to previously disease-free cacao farms. Two other ant species, *Camponotus acvapimensis* and *Pheidole megacephala,* also built tents, but these were made of soil and were usually situated near the ground. Nevertheless, if the soil contained propagules of the pathogen, rain splash could carry it to nearby pods (Evans 1973b).

Harry's work was confirmed a few years later by the British plant pathologist Alan Maddison, who demonstrated that tents built from debris by small ants could start outbreaks of black pod high up in cacao trees (above 5 m) (Maddison and Griffin 1981). The evidence supporting a role for ants, especially tent-building ants, in the epidemiology of black pod was accumulating.

Figure 5.5. Tent on a cacao pod formed by *Crematogaster striatula* ants. This is a carton tent, built from material in the litter layer, especially the epidermis of decaying pods; it is a potent vector of black pod. The scale insects visible on the pod belong to the genus *Stictococcus*. By permission of Harry Evans.

To prove the point, work led by Mike Griffin in Nigeria in the mid-1970s found that nearly 40% of black pod infections in their study plot were caused by ant activity, with the remaining infections resulting from rain splash (Taylor and Griffin 1981). And the story doesn't end there, because work published in 2018 by Akhmad Rizali and colleagues and carried out in Central Sulawesi in Indonesia showed that the greater the number of ant species accommodated by a cacao tree, the greater the incidence of black pod disease, presumably because more of the ant species were vectors of the pathogen (Rizali et al. 2018).

We've concentrated on ants, but they were not the only invertebrates to visit cacao pods suffering from black pod disease. In his research in Ghana in the 1970s, Harry found that a small dark-brown beetle, *Brachypeplus pilosellus,* and a long-legged fly, *Chaetonerius latifemur,* were common on pods throughout the year, but were most abundant during the wetter months. In fact, there seemed to be a succession of invertebrates on afflicted pods. Pods which were newly infected attracted the beetle and various flies, followed a little later by mollusks and millipedes. As the pods dried out and blackened further, other beetles took over. Harry's experiments demonstrated that the brown beetle and the long-legged fly could transport the black pod pathogen within the cacao tree and from tree to tree (Evans 1973b).

Of course, cacao pods are also attractive to vertebrates. Those responsible for damaging cacao range from elephants and wild cattle to squirrels and monkeys. In Trinidad and Tobago, orange-winged parrots and red-tailed squirrels feed on pods and destroy them in the process. To make matters worse, if they have feasted on infected pods, they can then transport the pathogen to healthy pods on other trees (Surujdeo-Maharaj et al. 2016, 238).

Sex and the Black Pod Pathogen

Although the black pod pathogen can get along quite nicely without sex, a celibate existence won't provide the genetic variation and diversity required to adapt to changes in its environment, either on or off the cacao tree. As with other Oomycetes, sex involves the getting together, and fusion, of male and female structures—antheridia and oogonia, respectively. Some *Phytophthoras* are self-fertile (known as homothallic) and others are self-sterile (heterothallic). In self-sterile species such as the black pod pathogens *P. palmivora* and *P. megakarya,* production of the sex structures requires the presence of two mating types, known as A1 and A2. Basically, a growth-regulating compound produced by one mating type stimulates the other mating type to produce sex structures. Sex occurs when a fertilization tube from the male antheridium penetrates the wall of the female oogonium and deposits its protoplasm within it. An oospore forms within the oogonium, developing its characteristically thick wall. The recombination of genetic information that results from the fusion of the two mating types can lead to the production of new races of the pathogen, some of which might be better adapted to infecting newer cacao varieties (Surujdeo-Maharaj et al. 2016, 221).

Although the two mating types A1 and A2 exist in *P. palmivora* and *P. megakarya*, they are not equally distributed. So, in *P. palmivora*, A2 is the dominant mating type across the cacao-growing world, apart from Jamaica, where the A1 mating type is more common, while in *P. megakarya*, A1 is the dominant mating type (Surujdeo-Maharaj et al. 2016, 228–229).

Phytophthora Rivalries in West Africa

We saw earlier that the major cause of black pod in West Africa is *P. megakarya*. It was reported in Nigeria in 1979 and since then has been recorded in Togo, Ghana, Côte d'Ivoire, and Cameroon. The year 1985 marked a turning point for black pod in Ghana. Until then, the only known cause of the disease was *P. palmivora* and that was bad enough. But things changed, because *P. megakarya* began ousting *P. palmivora* as the major cause of black pod in Ghana. It might not seem like much—after all, they are both *Phytophthoras* and the only difference is the species name. How bad could it be? In truth, bad, very bad, because *P. megakarya* is much more aggressive than *P. palmivora*. It starts producing spores earlier than *P. palmivora*, producing more of them and, combined with a substantial amount of the pathogen in the soil, ensures that more cacao pods are infected with *P. megakarya* each disease cycle. In fact, *P. megakarya* is so aggressive that it has displaced *P. palmivora* as the major cause of black pod in Nigeria and Cameroon (Bailey et al. 2016, 286). This has not occurred in Ghana, so far, where *P. megakarya* is still thought to be an invasive pathogen and has not yet become endemic in the west of the country. At present, both pathogens continue to cause substantial losses to the cacao crop in Ghana (Ali et al. 2016).

The continuing displacement of *P. palmivora* by *P. megakarya* in West Africa has occupied cacao pathologists for decades. The fact that *P. megakarya* produces lots of spores more rapidly than *P. palmivora* is only part of the puzzle and its completion will take some time. But thanks to the tools provided by the rapidly evolving field of genomics, new pieces are being added to the jigsaw, giving us a clearer picture of why the rivalry between these two black pod pathogens is playing out as it is.

P. palmivora can infect a wide range of host plants. In fact, it is known to cause disease in more than 1000 plant species. Contrast this with its rival, *P. megakarya*, which is not known to cause significant disease on any species other than the cacao tree (Bailey et al. 2016, 279). Since the cacao tree is native to South America, its transport to West Africa presented pathogens there, including *P. megakarya*, with a potential new host species. *P. megakarya* certainly took a liking to this South American import, with devastating consequences. In West Africa, it can cause yield losses of 30%–90% annually, and although losses from *P. palmivora* globally are 20%–30%, it has not caused major crop losses in West Africa in recent times (Ali et al. 2016). *P. megakarya* is clearly more damaging to cacao than its rival, but why should this be so? The answer is most likely complex and may be due, in part, to the ability of *P. megakarya* to survive the dry season in West Africa better than *P. palmivora*, and during the rainy season, it is spread more effectively by rain splash than its rival (Ali et al. 2016).

But part of the answer may lie in the recently published genome sequences for the two black pod pathogens (Ali et al. 2017). The work was undertaken by a global consortium of cacao researchers led by Bryan Bailey at the United States Department of Agriculture in Beltsville, Maryland. The results show that the evolutionary histories of the two species are very different, with *P. palmivora* duplicating its entire genome and *P. megakarya* selectively increasing the numbers of specific genes, many of which are responsible for virulence—the ability of a pathogen to infect a host and cause disease. It seems that duplicating its genome provided *P. palmivora* with the wherewithal to attack and live off many different host plants. In other words, *P. palmivora* is a generalist, a pathogen that is not specifically adapted to cacao, and maybe not to most of its hosts. This augmented genetic capacity has provided it with the ability to break down tough plant cell walls effectively, giving it access to the plants' nutrients and, in the process, enabling it to attack many different host plants without the need for adapting to specific hosts and all the time and effort that entails. Ultimately, since *P. palmivora* has a great many host plants to choose from, the selective pressure on it to adapt to the cacao tree and compete vigorously with *P. megakarya* might be low (Ali et al. 2017).

The evolutionary journey of *P. megakarya* has been very different. Adapting to a host, coevolving with it, means having to deal with its defense systems. Plants don't activate their defenses at the drop of a hat—that would be far too wasteful. Just imagine using up loads of precious energy and supplies getting your defenses ready to go and then no attacker turns up. No, it is far better to have an early warning system, something that can recognize an attacker and an impending attack before defenses are activated. Basically, plants have gate-keepers—proteins whose job it is to identify a genuine intruder, after which the signal is given to trigger defense. To get at a plant's nutrients, the pathogen must avoid recognition by the gatekeeper. So, over time, the pathogen evolves proteins that block the ability of the plant's gatekeeper to identify it, allowing the pathogen to get past them and into the plant's cells. These proteins allow the pathogen to become more virulent, that is, able to get into the plant, steal the goodies, and in doing so, causing disease in the plant. Of course, the plant is not going to take this lying down. Given time, it will evolve the means of preventing the pathogen's proteins from doing their sneaky job and, once again, it will be able to recognize the attacker. This game of cat and mouse is never-ending. So, increasing the numbers of virulence genes has enabled *P. megakarya* to adapt to the cacao tree rapidly, enabling it to effectively evade the tree's defenses to get at the nutrients it needs to complete its life cycle. But hang on a minute. If the interaction between the cacao tree and *P. megakarya* is less than 200 years old (Zhang and Motilal 2016), was there sufficient time for this adaptation to occur? Possibly not, and indeed, it has been suggested that the genetic changes allowing *P. megakarya* to be such a virulent pathogen on cacao occurred on other host plants, prior to its encounter with the cacao tree.

The purported center of diversity of *P. megakarya* lies on the border of Cameroon and Nigeria. A survey undertaken in an ancient primary forest in western Cameroon (Korup National Park) identified *Irvingia* sp. as a wild host of the pathogen, found in fallen fruits, most likely those belonging to the elephant mango, *I. gabonensis* (Holmes et al. 2003). *Irvingia* spp. are used in West Africa for trade and domestic purposes, and they are transplanted from the forest to cacao plantations to provide shade (Ndoye et al. 1998; Ayuk et al. 1999). Researchers reckon that collection of fallen fruit as food could spread *P. megakarya* from the forest to cacao plantations. In addition, the forest elephant eats fallen *Irvingia* fruit, providing another potential route for dispersing *P. megakarya* from forest to cacao plantation (Holmes et al. 2003).

Tackling Black Pod

Diseases caused by *Phytophthoras* are notoriously difficult to control. So too are diseases of tree crops. Combining the two, *Phytophthoras* and tree crops, just compounds the difficulty. Cacao trees can carry pods for much of the year, but the five-month period from the start of flowering through to harvesting is a time of great risk for cacao pods. This period usually coincides with the rainy season, which provides ample opportunity for the black pod pathogens to spread, by rain splash, for example. Spraying with a copper-based fungicide is the standard means of trying to control black pod, although newer systemic fungicides are available (Surujdeo-Maharaj et al. 2016, 240–242). Copper has a long history of use as a fungicide, starting way back in 1807 with the Frenchman Bénédict Prévost, when he discovered that wheat seeds, placed in a copper colander and steeped in a fragrant mixture of lime and sheep urine, produced a crop free of smut disease (Money 2007, 106). Although others attributed the smut-control to the lime, Prévost showed that it was copper leaching from the colander that provided the best protection against the smut. Copper has been used as a fungicide ever since, but continuous use leads to its accumulation in soils and plants, with consequences for soil fertility and the environment. In fact, use of copper fungicides to control black pod is thought to be responsible for copper contamination of soils in Nigerian cacao farms, and copper accumulation in leaves and pods of cacao in Ghana (Aikpokpodion et al. 2010; Addo-Fordjour et al. 2013).

Chemical control is rarely completely effective, especially since efficacy depends on the stage of pod development, prevailing weather conditions, and getting good pod coverage. And yet, despite these difficulties and the cost of the chemicals, this is a preferred option for many cacao farmers (Surujdeo-Maharaj et al. 2016, 240). In Ghana, under a national program to control black pod caused by *P. megakarya,* copper fungicides are applied three times during the rainy season (Ali et al. 2016). This places an enormous strain on resource-poor farmers. Since *P. palmivora* is less aggressive and damaging than *P. megakarya,* it is advisable to know which of the two pathogens is causing the black pod, before the decision is made to spray. This is easier said than done, since it requires a surefire way of distinguishing between the two. Luckily, researchers have recently developed a quick and simple means of identifying the two *Phytophthoras,* based on genetic differences between them (Ali et al. 2016). The test uses polymerase chain reaction (PCR) to produce millions of copies of a small segment of DNA that is specific to individual species of *Phytophthora.* This brilliant technique revolutionized the study of DNA so profoundly that its creator, Kary Mullis, was awarded the 1993 Nobel Prize in Chemistry. Anyway, getting back to black pod—it is hoped that this new PCR test will enable farm advisers to determine which *Phytophthora* is causing black pod, allowing them to devise an appropriate control strategy such as spray or don't spray.

Applying chemicals is but one approach to tackling crop diseases. An important part of any attempts to control crop disease is hygiene (Surujdeo-Maharaj et al. 2016, 238–240). The basic idea here is to promote healthy growth of the cacao trees, while making life as difficult as possible for the pathogen. For the black pod pathogens, this can be done by decreasing the relative humidity within the canopy of the cacao trees. Zoospores of the pathogen require water to germinate and swim around as they try to locate an appropriate point to infect the plant; reducing humidity within the canopy creates a less favorable environment for the spores, thereby reducing infection. Reducing shade, weeding regularly, and pruning the trees helps to increase air flow through the canopy, reducing humidity in the process.

It stands to reason that leaving infected pods on the tree is not a good idea, as it provides a ready source of pathogen inoculum to start further infections. Infected pods can be completely colonized by the pathogen in as little as four days, with each infected pod capable of producing some 4 million sporangia (Bailey et al. 2016, 294). Pods should be harvested often to remove any pods showing signs of black pod. This goes for mummified pods, too, which can act as a source of black pod inoculum for several years. Discarded pods and husks should then be taken away and not simply left at the base of the tree to act as a source of pathogen inoculum, to splash onto pods on nearby trees, or carried by insects to other trees on the farm.

Enlisting Microbial Help

Controlling plant diseases has never been easy. Today, with increasing pressure to reduce pesticide use and the difficulty of producing plant resistance that will truly last, there is a great need for additional approaches to tackle diseases. One approach that has received considerable attention over the past few decades is biological control—using microbes to fight plant disease. It might sound counterintuitive, but it's not. Microbes that cause plant disease are food for other microbes, so why not make use of them?

One group of fungi that love to feast on other fungi belongs to the genus *Trichoderma*. Many species of these feisty fungi have been studied with a view to using them to control crop diseases. One member of this group with an appetite for the black pod pathogen *P. megakarya* is *T. asperellum*. For the sake of brevity, let's call it *Tasp*. One isolate of *Tasp* was tested in field trials in Cameroon and acquitted itself well, reducing black pod by 66% by year 2 of the study. But then the plague of biocontrol work kicked in—inconsistency. In subsequent trials, the results were variable, and researchers suggested that the way the fungus was formulated was part of the problem. And sure enough, when a new oil-based formulation was tried, disease control was better and more consistent. It seems that the beneficial effects of the formulation were due, first, to its constituents forming a protective barrier around the cacao pod, thereby making pathogen infection more difficult, and second, to protecting *Tasp* spores against the vagaries of the climate (ten Hoopen and Krauss 2016, 542).

A huge number of microbes live within plant tissues without causing damage or symptoms. These endophytes, as they are called, include a great many fungi and, incredibly, they have been found in healthy tissues of all plants examined to date (Arnold et al. 2003). Working in a lowland forest in Panama, researchers found that the tissues of the cacao tree harbor a great diversity of fungal endophytes, all eking out a living in the spaces between the plant cells by absorbing whatever nutrients leaked out of the cells. Although these "silent" lodgers were not harming the tree, the researchers wondered whether they provided any benefit to the plant. So, they grew cacao seedlings in a sterile environment to produce plants containing none of the endophytes they would usually harbor. They then examined whether these plants were more likely to become infected with black pod and whether reintroducing the endophytes would protect the young plants. The results were striking—plants without the fungal endophytes suffered badly from black pod, whereas the plants that had been reunited with their fungal lodgers successfully warded off the pathogen. The endophyte-induced protection was greater in mature cacao leaves than young leaves, which is just as well, since mature leaves lose a lot of their intrinsic chemical protection as they age (Arnold et al. 2003).

It's one thing to discover the role of fungal endophytes by introducing them to plants grown in a sterile environment, but what happens naturally, in the field? This question was examined by researchers in Panama, who looked at whether varying the leaf litter around cacao plants altered the microbiome (the total population of microbes) in the seedlings and, in turn, whether that had any effect on the ability of the seedlings to defend themselves against black pod. They found that manipulating the composition and location of leaf litter within the forest canopy changed the cacao microbiome. In particular, exposing cacao seedlings to leaf litter from healthy adult plants enriched their microbiome with a fungal endophyte, *Colletotrichum tropicale*. The researchers knew that this fungal lodger had been shown to enhance the ability of cacao plants to resist pathogens by activating defenses. And the effect here was the same—growing the seedlings in leaf litter from healthy adult trees reduced damage from black pod (Christian et al. 2017).

Resisting Black Pod

Clearly, *Phytophthoras* are difficult to control, and those species responsible for black pod are no exception. But cacao, like all plants, possesses a wide range of defenses that it can use against pathogens and other attackers. Some of these defenses are always present; think of surface waxes on cacao pods. Researchers in Ghana found that cacao genotypes with more wax on their pods exhibited greater resistance to the black pod pathogen than cacao genotypes with less waxy pods. Surface waxes add an extra layer of thickness, a barrier if you like, making it more difficult for pathogens to get into the host. They can also contain chemicals that disrupt normal development of the pathogen, making it impossible for them to penetrate the plant surface. Waxy surfaces are also hydrophobic, causing water to form droplets which either run off or evaporate quickly. An extra-waxy and drier pod surface would be a hostile environment for moisture-loving *Phytophthora* spores, with many dying before they can start the process of penetrating the pod surface (Nyadanu et al. 2012). More recent research in Ghana identified cacao genotypes that would be suitable for breeding improved varieties with a thicker wax coating on their pods (Nyadanu et al. 2019).

As we saw earlier, both *Phytophthoras* can enter cacao pods via stomata, although *P. palmivora* is more likely to use this mode of entry than *P. megakarya*. Researchers in the Cocoa Research Unit at the University of the West Indies in Trinidad found that resistance to black pod caused by *P. palmivora* was related to the frequency of stomata on the pod surface as well as to their size. So, cacao clones with fewer stomata and/or smaller stomatal apertures were more resistant to the pathogen than clones with lots of stomata or stomata with larger apertures, leading the researchers to suggest that selection for these pod characteristics might be useful in programs aimed at breeding for black pod resistance (Iwaro et al. 1997).

But what happens if the black pod pathogen gets through the outer surface of the pod, or through the stomata, or is lucky and enters via a wound? Does this mean curtains for the pod? What can the pod do at this stage? If external barriers and defenses have failed, the pod can turn to its chemical enforcers. One group of chemicals the pod will have at its disposal is phenolics, a diverse and widespread group of compounds that will be more familiar to us than we might realize. Some are constituents of essentials oils, responsible for the wonderful scent and flavor of many herbs. A good example is thymol, the major component of the essential oil produced by thyme, great in cooking, but also used in mouthwashes. Then there are the anthocyanins, phenolic compounds responsible for the red, blue, and purple colors of many flowers, fruits and vegetables, some of which are believed to account for the reported health benefits of fruits such as blueberries and goji berries. But there is more to phenolic compounds than smell, taste, and color. If you are a microbial pathogen, they can really spoil your day, for many of the phenolics found in plants have a defensive role—they can kill microbes (Walters 2017).

As we've seen in earlier chapters, cacao is full of phenolic compounds, and several studies have implicated them in the ability of cacao leaves to resist infection by the black pod pathogens. Researchers in Cameroon found that the ability of two Trinitario clones to successfully resist infection by *P. megakarya* was associated with a high phenolic content. When these two clones were crossed, their progeny were even more resistant to black pod, and this resistance was also linked to a greater phenolic content (Djocgoue et al. 2007). The major phenolic compound in cacao leaves was apigenin, accounting for nearly half the total phenolics. Apigenin, together with another phenolic, luteolin, accumulated in leaves resisting attack by the black pod pathogen. So, what happens in pods? After all, the disease is called black pod. Well, phenolics are also linked to enhanced resistance to *P. megakarya* in cacao pods, and attempts by the pathogen to invade pods of resistant clones leads to accumulation of phenolic compounds (Boudjeko et al. 2007).

Whatever mechanisms are used by the cacao tree to resist attack by black pod pathogens, they are all regulated by the tree's genes. Individual genes or groups of genes will be responsible for different defenses. So, the phenolics that are important in fending off the black pod pathogen will be made by specific genes. Now, resistance comes in two main types in plants: that controlled by one or a few genes (known as monogenic resistance), and that controlled by many genes (known as polygenic resistance). We saw earlier that pathogens and plants are engaged in a never-ending game of tit for tat: the plant develops the means to detect an invading pathogen and over time the pathogen develops a way of disguising itself, thereby avoiding detection, with this cycle repeated time and again throughout evolutionary history. The ability of the plant to detect the pathogen is governed by one or just a few genes; that is, the resistance is monogenic. When the pathogen evolves the means of avoiding detection, a new "race" of the pathogen is born. So, eventually, you end up with many races of the pathogen, each new race having evolved to overcome the latest attempt by the plant to detect it. But monogenic resistance is an all-or-nothing affair: pathogens can evolve the means of overcoming this type of resistance rapidly, because the pathogen only needs to deal with one or just a few genes. By contrast, pathogens find it much more difficult to overcome polygenic resistance, because here, it needs to overcome many genes. This type of resistance is much longer lasting than monogenic resistance (Walters 2011, 206–208; Walters 2017, 156–158).

Where does black pod fit in all of this? As we've come to expect with black pod, nothing is straightforward. Although some workers suggest that resistance to black pod caused by *P. palmivora* is polygenic (Micheli et al. 2010; Barreto et al. 2015), others have suggested that some sources of resistance are monogenic. So, research conducted in Brazil in the 1990s suggested that resistance to black pod caused by *P. palmivora, P. citrophthora,* or *P. capsici* appeared to be monogenic (Barreto et al. 2015). This finding was confirmed some 20 years later, when a group of Brazilian researchers obtained strong evidence that resistance to the black pod pathogens in Brazil is indeed monogenic (Barreto et al. 2015). It would seem, therefore, that both sources of resistance exist.

Ultimately, sustainable control of black pod requires the use of cacao clones resistant to the pathogen, and this is where plant breeders come in. Generating new cacao clones for use by farmers begins with collecting and characterizing cacao germplasm. It is likely that some species of *Phytophthora,* perhaps *P. palmivora,* coevolved with the cacao tree in South America and it stands to reason that a good place to look for resistance to them is in the tree's center of origin, thought to be the upper Amazon. Surveys of wild cacao trees in southeastern French Guiana between 1985 and 1995 identified trees with good levels of resistance to black pod. Trees in this collection, belonging to the 'Guiana' genetic group (see chapter 3), constitute an important source of resistance to black pod caused by *P. palmivora, P. megakarya,* and *P. capsici* (Lachenaud et al. 2015; Thevenin et al. 2012) and could be incorporated into cacao breeding programs. Subsequent research carried out in Brazil and published in 2015 identified ten genotypes with resistance to the three main black pod pathogens (Lachenaud et al. 2015). These genotypes will be important in selecting genes for black pod resistance for use in breeding programs.

As with other perennial tree crops, traditional cacao breeding is very slow and takes a long time. A single selection cycle takes more than 10 years and two or more cycles are often needed before a new variety can be released to farmers. So, how likely is it that truly effective resistance to black pod can be attained? Some workers think it is unachievable (Bartley 1986; Lachenaud et al. 2015). Why? First, because black pod in different regions can be caused by several *Phytophthora* species with a wide range of aggressiveness. Then there is the lack of genetic variability available in cacao breeding programs, and together, they stack the odds against the plant breeder. Nevertheless, there is hope. Cacao scientists have powerful new genomic technologies at their disposal, and the cacao genome has been sequenced. This should help cacao breeders identify resistance genes for use in breeding programs. It is hoped that the incorporation of new marker-assisted selection techniques, where specific genes for disease resistance are tracked, will speed up the development of cacao materials with significant resistance to black pod rot. It is important to remember, however, that wherever cacao is grown, the tree and the black pod pathogens continue to evolve together, so resistance is never going to be everlasting.

Researchers reckon that in Ghana, areas previously affected by *P. palmivora* suffered a staggering 70% reduction in cacao production following invasion by *P. megakarya* (Holmes et al. 2003). Cacao-growing regions of the world not yet afflicted by *P. megakarya*, but which are currently affected by other devastating cacao diseases (see the following chapters), could do without another disease burden. The need for vigilance is clear.

6

Witches' Broom

In 1783, the Brazilian-born naturalist Alexandre Rodrigues Ferreira held a position at the Museum of Ajuda in Lisbon. Later that year, the 27-year-old Ferreira found himself traveling back to Brazil to start an expedition into the Amazon basin, on the orders of Queen Maria I of Portugal. This region of Portugal's colony remained virtually unexplored, and Ferreira's instruction was to undertake a "philosophical" voyage through the vast region, as prelude to economic exploration. He was two years into what turned out to be a nine-year expedition, when he came across some strange growths on cultivated cacao trees. He described the growths as *lagartão*, or big lizard, and in so doing, provided the first report of what subsequently came to be known as witches' broom disease of cacao (Wood and Lass 1985, 282; Money 2007, 73). Just over one hundred years later, a widespread outbreak of witches' broom was reported in Suriname, and within ten years, the cacao plantations established by Dutch planters in the eighteenth century were devastated. The disease spread rapidly, reaching Guyana in 1906, Ecuador in 1918, and Trinidad in 1928, followed by Colombia the following year and the island of Grenada in 1948 (Meinhardt et al. 2008). Everywhere it struck, the effects on cacao production were catastrophic. Cacao crops were destroyed entirely in Guyana, and it took decades for Ecuador to regain its position as a major cacao producer.

Brazil escaped the ravages of witches' broom for a long time. The cacao crop was cultivated largely in Bahia, on the coast, well away from the Amazon basin where the tree originated and where it had coevolved with the witches' broom pathogen. But the advent of the Trans-Amazonian Highway changed everything (Evans 2016, 146). Its construction in the 1970s led to a huge expansion in cacao cultivation in Brazil's rainforest states, which were so productive that the country was on course to become the world's leading exporter of cacao beans. But it was asking too much to expect the sudden appearance of vast areas of its host plant to go unaffected by the witches' broom pathogen. And sure enough, the trees became infected, and despite attempts by growers to eradicate the pathogen, by the late 1980s, a third of the cacao crop in the state of Rondônia was lost every year to witches' broom. But at least the cacao being produced in Bahia, some 2575 km away, was safe from the clutches of witches' broom. After all, its spores were hardly the Bear Grylls of the fungal world and were unlikely to survive more than 24 hours without gaining access to their host—if they did not infect a cacao plant within that period, they were goners. What's more, there was a large, cacao-free area—the arid to semiarid, spiny forests of the Cerrado and Caatinga ecosystems—lying between the rainforest plantations to the west and the coastal plantations to the east, more than enough to prevent the spread of the pathogen's less than impressive spores. Since the only way for the pathogen to cross this barrier was on the plant, quarantine measures were introduced to prevent the transport of infected plants from the rainforest to the coast (Evans 2016, 146).

All was well until mid-1989, when the disease was discovered on a farm close to the town of Uruçuca. Although initial attempts to contain the disease were successful, eventually the team responsible took the decision to burn everything—there was, quite simply, too much at stake. Then, just two months later, witches' broom was discovered on a cacao farm 119 km south of Uruçuca. But anyone thinking that the pathogen causing the disease in Uruçuca had just moved a few miles south was proved wrong. Using genetic fingerprinting techniques, researchers found that the two outbreaks in Bahia were caused by different strains of the pathogen, matching those present in Rondônia. Rather than witches' broom entering Bahia once, it had been introduced twice (Andebrhan et al. 1999; Money 2007, 73), and since the pathogen could not have reached Bahia unaided, its appearance there was attributed to human intervention. There has been speculation that these introductions were deliberate (Gadsby 2002; Smallman and Brown 2012), since the first "introduction" took place along a road and the second along a river, and in both cases, instead of the infection appearing at the edge of the farms, they sprang up in the middle. Although this does not prove that the introductions were deliberate, a news article published in 2006 in Brazil contained a confession from a man who claimed to have introduced the disease at the two locations. The motive was to destroy the economy and break the hold of the powerful right-wing cacao barons in favor of the opposition, left-wing party (Evans 2016, 148). But questions were raised about the man's credibility and a subsequent public hearing ruled that there was insufficient evidence to confirm the veracity of the story. Today, although there is still debate about whether these introductions were acts of bioterrorism, some commentators consider them to represent one of the first acts of agro- or bioterrorism for political as opposed to financial gain (Evans 2016, 148).

We may never find the truth of what really happened in Bahia in 1989, but one thing is clear—the effects were calamitous and far-reaching. Cacao production in Bahia plummeted by 60% within four years, and by 1994, cacao production in Brazil dropped from 400,000 to 100,000 tonnes. As a result, 200,000 people lost their jobs in the cacao industry in Bahia and considerably more were affected. Ruined farmers fled the cacao-growing areas for cities, which saw increases in homelessness and crime (Money 2007, 73).

This was a disaster for Brazil, but across the Atlantic in West Africa, cacao growers in Ghana saw things differently. Once in competition with Brazil for the Number 2 position in the cacao-producers league table, Ghana watched as Brazil's output plummeted. By 2016, Ghana occupied second position in the table, with an output of 835,000 tonnes a year, while Brazil, although lower down the league table, produced a still respectable 256,000 tonnes of cacao beans (World Atlas n.d.). The bulk of the world's cacao is now produced in West Africa and Indonesia, countries where the cacao tree was introduced and where it has thrived, free from witches' broom. Let's hope it stays that way.

What's in a Name?

Scientific interest in witches' broom started in the late 1800s when the disease was laying waste to cacao plantations in what was then Dutch Guiana (now Suriname), but the identity of the causal organism caused considerable confusion until well into the twentieth century. These early studies in Suriname resulted in the first confirmed description of the disease and led to its Germanic common name, *krülloten*, which translates as "witch-broom" (van Hall and Drost 1909; Evans 2012). Over the years, it was placed in four different genera of fungi until 1907, when it was decided that the agent responsible was the fungus *Colletotrichum luxificum* (Evans et al. 2013). Not everyone accepted this classification, however, including James B. Rorer, an American mycologist based in Trinidad. After isolating the fungus from brooms and pods in Suriname in 1913, he concluded that it resembled a basidiomycete, the class of fungi that form mushrooms (Rorer 1913; Evans et al. 2013). The witches' broom fungus does indeed produce mushrooms—small, pink to crimson-colored fruiting bodies that emerge from the necrotic brooms. A couple of years after Rorer's work, the Swiss mycologist Gerold Stahel, working in Suriname, discovered that if he suspended dead witches' brooms bear-

Figure 6.1. Mushrooms (basidiocarps) of the witches' broom fungus, *Moniliophthora perniciosa*. By permission of Harry Evans.

Figure 6.2. Underside of a mushroom (basidiocarp) of the witches' broom fungus, *Moniliophthora perniciosa*. By permission of Harry Evans.

ing mushrooms over cacao seedlings, they eventually developed symptoms typical of the disease (Stahel 1915). Thinking that the fruiting bodies looked like the tiny structures produced by many species of *Marasmius,* he named it *Marasmius perniciosus.*

The situation remained like this for the best part of 30 years, when the fungus was transferred to the genus *Crinipellis,* as *C. perniciosa,* by the mycologist Rolf Singer in 1942. When I first came across witches' broom disease as an undergraduate in the mid-1970s, I knew it as *Crinipellis perniciosa,* although it was often referred to as *Marasmius perniciosus.* Some workers even went so far as to suggest that the fungus should continue to go by the name of *M. perniciosus,* since everyone knew it as such (Baker and Holliday 1957; Evans et al. 2013). Harry Evans tells the story of seeing Singer, at an international conference in Brazil in 1978, growing increasingly agitated as talk after talk referred to the witches' broom pathogen as *Marasmius perniciosus,* despite Singer's having talked earlier at the conference of the pathogen's position within the genus *Crinipellis* (Evans 2016, 150). Eventually, in 2005, DNA sequencing studies demonstrated that the witches' broom fungus is very closely related to another disease-causing fungus on cacao—the frosty pod rot pathogen, *Moniliopthora roreri,* which we will be looking at in the next chapter. In fact, the witches' broom fungus is only distantly related to other *Crinipellis* fungi. So, at long last, the witches' broom pathogen had its true identity—*Moniliophthora perniciosa.* It remains to be seen whether this is the final word on the identity of the witches' broom fungus.

Sorting out the identity of the witches' broom fungus has not, however, put an end to the intrigue that follows it around. *M. perniciosa* might be best known for infecting cacao, but it can also infect other plants. Three biotypes of the fungus are currently known, each adapted to living on a different group of host plants. One biotype (C) infects the cacao tree and the closely related genus *Herrania,* another biotype (S) infects solanaceous plants, while the third biotype (L) infects liana vines of the Bignoniaceae family of plants in western Ecuador. Unlike the C and S biotypes, which induce witches' broom symptoms in their hosts, the L biotype does not induce symptoms in its vine hosts; instead, it seems to live unobtrusively, as an endophyte (Evans 1978; Griffith and Hedger 1994). Perhaps unsurprisingly, the C and S biotypes are genetically similar, but even so, the C biotype cannot infect solanaceous plants and the S biotype can't infect cacao. By contrast, the L biotype is genetically divergent from the other two biotypes, exhibiting higher levels of genetic variability. This reflects the fact that the L biotype needs to cross with a different mating type to produce the mushrooms and hence the spores required to infect its host. The C and S biotypes don't need to go to the trouble of finding a different mating type—they can do it themselves. At a stroke, the business of mating and producing the spore-containing mushrooms becomes more convenient and straightforward. Researchers reckon that this favors their pathogenic lifestyles because spreading and colonizing new host plants does not require finding a mating partner at their new location (Mondego et al. 2016, 186).

Jekyll and Hyde

The tiny pink mushrooms that emerge from the dead cacao brooms harbor the next generation of the pathogen. On the underside of these caps, the gills bear millions of miniscule spores known as basidiospores, clouds of which erupt from the mushrooms, mostly at night, to ferry the fungus to other cacao plants. Researchers found that an increase in relative humidity coupled with a drop in temperature favors release of the spores. The spores are, in fact, programmed for release at night, thereby escaping the high temperatures and lower humidity of the day, which would kill the thin-walled spores (Frias and Purdy 1991; Evans 2016, 167). Providing the humidity is high, these microscopic spores can infect meristematic tissues of the cacao tree—shoots, flowers and young developing fruits. The spores might look fragile and inconsequential, but they can penetrate the epidermis directly, as well as via stomata. Entering the plant via stomata is easier if they are open, but they close at night, just as the witches' broom fungus is on the hunt for suitable entry points. This, however, appears to be but a minor inconvenience for the fungus, since it can secrete oxalic acid, which is known to prevent stomata from closing at night (do Rio et al. 2008). Welcome to the fungal world of breaking and entering.

Like the black pod pathogen, the witches' broom fungus is a hemibiotroph

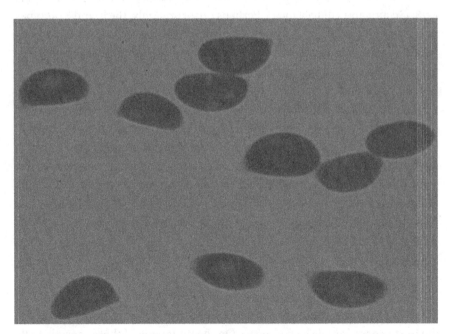

Figure 6.3. Basidiospores produced by the mushrooms (basidiocarps) of the witches' broom fungus, *Moniliophthora perniciosa*. By permission of Harry Evans.

(Meinhardt et al. 2008). Initially, the hyphae that snake their way between the cells of the cocoa tissues are biotrophic, obtaining their sustenance without killing the plant cells. Essentially, they live off the stuff that leaks out of the plant cells. This might not seem like much of a meal, but the sugars and amino acids that end up in the spaces between the cells—the apoplast—are just enough to keep the pathogen going. It is possible that the fungus can alter the permeability of the plant cells, causing nutrients to leak out in greater quantities, which would provide more fuel for the growing but nonpaying tenant. Even so, the fungus can only grow slowly on the meager rations available between the plant cells. But unlike the impatient black pod pathogen, whose benign phase is short-lived, lasting just a few hours, with the witches' broom fungus, the biotrophic phase can last for up to three months. Then, after several months of gentle parasitism (if there can be such a thing), the fungus turns into a monster, killing the cells of its host, before feasting on the dead remains—charming. This killing spree—the necrotrophic phase—leaves the cacao shoots brittle and dry, and before long, the pink mushrooms form and the cycle starts again (Meinhardt et al. 2008).

Now, you might well be wondering whether the pilfering of what little food exists between cells in the cacao tissues can lead to the grotesque malformations that constitute the witches' brooms. Put simply—no. Although this is still a poorly understood area, it seems likely that following pathogen infection, the balance of plant hormones is altered, leading to various changes, including huge increases in the numbers of plant cells (known as hyperplasia) and the sizes of the individual cells (hypertrophy), as well as a loss of apical dominance. Plants tend to grow upward because they have a bud at the top of the stem—the apical bud. Although plants produce buds in the axils of leaves, these don't grow initially because of the dominant influence of the apical bud. In witches' broom disease, loss of apical dominance means that the axillary buds grow out and side shoots proliferate, leading to the formation of the sinuous, lizardlike brooms from which the disease gets its name (Teixeira et al. 2015). The changes in hormonal balance within the cacao tissues are responsible for more than the development of the green brooms—they are also likely to suppress the plant's defenses, allowing the fungus to grow slowly between the plant cells in its easygoing biotrophic state.

The curious among you might be wondering what happens to change the witches' broom pathogen from a relatively gentle biotroph to an all-destroying necrotroph. In 1980, the ever-insightful Harry Evans suggested that an unstable molecule produced by the cacao tissues was responsible for keep-

Figure 6.4. Cacao seedling infected with the witches' broom fungus. In infected seedlings, loss of apical dominance means that the axillary buds grow out and side shoots proliferate, leading to the formation of the sinuous, lizardlike brooms from which the disease gets its name. By permission of Harry Evans.

ing the fungus in its biotrophic phase (Evans 1980). If this molecule was not around, the fungus would switch quickly into its necrotrophic phase. So, what might this unstable molecule proposed by Harry be? Well, in 2012, researchers discovered that the elusive molecule is likely nitric oxide, a chemical messenger found in animals as well as plants (Thomazella et al. 2012). It seems that the nitric oxide produced by the plant regulates the way the fungus produces energy for its growth and development. Tiny organelles called mitochondria are the powerhouses of all cells, and they can produce energy in two ways: one yields lots of energy, and another producing considerably less. The nitric oxide keeps the mitochondria in the witches' broom fungus in the low energy production mode which, in turn, helps to keep the fungus in the slower-growing, less-demanding biotrophic phase. But it's hard to keep going when food is running out, and eventually the onset of starvation prompts the fungus to start killing plant cells to liberate the food locked within them. Dead cells don't produce nitric acid (or much else for that matter), the absence of which switches the fungal mitochondria to the higher-yielding mode of energy manufacture and, in so doing, signals the end of the biotrophic phase and the start of the necrotrophic phase (Thomazella et al. 2012).

The fungus in its new necrotrophic guise starts growing rapidly on the newly available food, but completing its life cycle requires yet another makeover. All the death and destruction transform its living quarters from green tissues to dead, dry brooms. It needs to find food in this seemingly unpromising environment, but there is another trick up its sleeve, for having switched from biotrophy to necrotrophy, it now switches to saprotrophy. This remarkable fungus now produces enzymes to degrade wood, as do those fungi that live exclusively on the stuff. But hang on—if other fungi are specialist wood eaters, how can the witches' broom fungus compete? By killing off its competitors, of course! The ever-resourceful fungus produces several proteins whose job it is to kill off other fungi trying to muscle in on the dry brooms. It also helps that the dry brooms don't fall off the cacao tree, thereby preventing contact with competing microbes in the leaf-litter (Mondego et al. 2016).

The ability of the witches' broom fungus to cause disease is exceptional among its close relatives, the majority of which, being saprotrophs, live quietly off dead things. So, how did it turn to the dark side and take up with the pathogen crowd? It seems that in the dim and distant past, an early ancestor of the witches' broom fungus obtained a few genes from some of its neighbors, who just happened to be relatives of the black pod pathogen, belonging to the genus *Phytophthora*. Among the genes they acquired by horizontal gene transfer from their pathogenic pals was one that makes a protein that kills plant cells (it's called a NEP protein, where NEP stands for Necrosis and Ethylene-inducing Protein). Today, the witches' broom fungus uses the NEP protein to help it in its switch from mild-mannered biotroph to nasty necrotrophy. What better way to "prepare" your host for your lifestyle switch than by killing its cells (Tiburcio et al. 2010).

Figure 6.5. Symptoms of pod infection by the witches' broom fungus. By permission of Harry Evans.

So, eventually, after three changes of lifestyle—biotrophy, necrotrophy, and saprotrophy—the fungus is now ready to produce its pink mushrooms and propel itself into the next generation. Once liberated, the spores don't have long to find a new host plant. It's reckoned that they are viable for about 12 hours, although they could last longer under overcast conditions. The fragile agents of doom are borne on the wind, although the distances traveled are unlikely to be greater than 16 km. Armed with this knowledge, it is easy to see why spread of witches' broom from Rondônia across 2575 km of mountains and semiarid forests to Bahia on the Atlantic coast was thought to be near impossible (Evans 2016, 167).

It's Not All about the Broom

The brooms from which the disease takes its name are certainly noticeable, but they probably reflect a small proportion of infection events that occur on cocoa trees. Evans reckons that most infections are restricted to swellings on green shoots or the petiole. These hidden infections are easily missed but are important if farmers are to manage the disease successfully (Evans 2016, 166). Although unpollinated flowers can become infected and give rise to abnormally developed pods, more serious is direct penetration of the young pod or cherelle. Such pods fail to develop properly and don't reach maturity, resulting in a total loss of beans. Looking inside the pods is a plant pathologist's delight but a farmer's nightmare, with the contents ranging from a mucilaginous mess to a dry, compacted mass of useless beans (Evans 2016, 164). Farmers can find it difficult to distinguish witches' broom pod infections from those caused by black pod or frosty pod rot, although with witches' broom infection, necrosis of the pod is delayed by two or three months, unlike black pod infection, where death of pod tissues takes just a few days. Frosty pod rot also causes brown lesions on pods, but these tend to become covered with the mycelium and spores of the frosty pod rot fungus, whereas no such fungal growth occurs on the dark brown lesions in pods infected with the witches' broom pathogen (Evans 1981).

A Clean Sweep: Tackling Witches' Broom

Good hygiene is important in controlling any disease, and one way of achieving this is to remove anything that can act as a source of the pathogen. One obvious source of witches' broom inoculum is the infected broom itself, so

Figure 6.6. Mummified pods (chirimoyas) on a cacao tree caused by infection of the pods by the witches' broom fungus. By permission of Harry Evans.

its removal will clearly reduce the amount of fungus able to start new infections. This stands to reason, since brooms on the tree produce three times more spore-bearing mushrooms than brooms removed and left on the ground. Pruned brooms on the ground decompose rapidly, which reduces their mushroom-producing potential. So, although it might be tempting to chop off the brooms and leave them on the ground, it is far better to take them away and burn them (Wood and Lass 1985, 290–293).

Sometimes observations made by farmers can help them to better manage witches' broom on their farms. A good example is the Brazilian farmer who, going against the traditions established for managing cacao in Bahia, moved pruning his trees from January–March to September–November. His rationale was simple—the peak of spore production by the witches' broom fungus is

January–March. Pruning during this period is followed by a flush of new cacao growth, providing lots of juicy new meristems for it to infect. By delaying pruning to later in the year, flushes of new cacao shoots are not emerging into an environment laden with spores of the witches' broom fungus. Cacao farms adopting this management practice drastically reduced incidence of witches' broom (Mondego et al. 2016).

The cacao tree's natural habitat is the understory of the rainforest, where it has evolved to grow in the shade of considerably larger trees. Cultivating cacao under heavy shade leads to slow growth, which means that new shoots and flowers are produced at lower rates. This makes for an unproductive crop, but the upside is that there is less soft, susceptible new cacao growth for the witches' broom fungus to take advantage of. Cacao grown without shade grows much more quickly and is very productive, but there is a cost—the availability of lots of lovely new growth, just right for the frightful fungus. But shade has other beneficial effects too. For example, it lessens fluctuations in temperature and humidity within the canopy of cacao trees, and since maximum spore production requires alternating periods of drying and wetting, shading can reduce the amount of inoculum produced by the fungus. And there's more—in unshaded cacao, infected shoots dry out quickly, which, in turn, reduces colonization by secondary organisms. This gives free rein to the witches' broom fungus, which can produce its mushrooms with their deadly cargo in greater numbers, over a longer period (Evans 1981; Wood and Lass 1985, 290–293).

The Farmer and the Microbe Can Be Friends

In the quest to control the witches' broom menace, the farmer does not have to do it alone. He can turn to that most unexpected of allies, the microbe, because although it might seem like it, not all microbes are nasty. In fact, some are downright friendly. Take the fungus that goes by the name of *Trichoderma stromaticum*, or Ts as we shall affectionately call it. Ts is an endophyte of cacao, meaning it grows within the tissues of the cocoa tree without causing any harm. Having someone in your home that causes no harm is a good start, but it's even better if the lodger performs some useful function too. In the case of Ts, its forte is giving the witches' broom fungus a hard time, because Ts is actually an effective mycoparasite of the plundering pathogen. Ts colonizes dead brooms and old infected cacao pods, feeding on the witches' broom fungus. This fungus-based feast reduces the production of witches' broom mushrooms, thereby lowering inoculum levels (De Souza et al. 2008; ten Hoopen and Krauss 2016, 544). But when Ts was put to the test over three years in the field, it did not protect cacao pods from witches' broom. It redeemed itself, however, when used alternately with copper hydroxide, reducing disease incidence by up to 60% and increasing pod yield per tree. In fact, Ts is available for use by cacao farmers in Bahia, where it goes by the name Tricovab. This biocontrol agent is the only one currently used on a large scale in Brazil, probably because it is produced and distributed by CEPLAC, the Brazilian government's Executive Commission for Cacao Farm Planning (ten Hoopen and Krauss 2016, 551). There is huge potential for using biological control against cacao diseases, but, as things stand at present, the lack of consistent disease control, together with the difficulties in registering new products, and the cost, is off-putting to the industry and cacao farmers.

Resistance to Witches' Broom

As we've seen before for black pod, use of fungicides to control witches' broom is impractical, though it is used routinely on developing pods. In an ideal world, the best approach to controlling witches' broom is the use of resistant varieties. Scientists began searching for resistance to witches' broom in Ecuador in the 1920s, followed by expeditions to the Amazon in 1937 and 1942 to collect disease-free material from semicultivated and wild cacao trees. Plants collected during this expedition were sent to Trinidad to be evaluated, leading to the identification of two Peruvian accessions with good resistance to the fungus. But these accessions had poor agronomic qualities and could not be cultivated. Instead, they were used to develop hybrids, such as the Trinidad Selected Hybrids and the Trinidad Selected Amazons, which have been used successfully in Trinidad and Brazil to reduce witches' broom (de Albuquerque et al. 2010). But resistance based on these hybrids did not last, with the witches' broom pathogen overcoming it in Ecuador, Brazil, and Peru. The search for resistance to witches' broom never stops, and in 2010, Brazilian researchers identified new sources of resistance in material collected from the Brazilian Amazon. Plants showing particular promise were collected from distinct river basins in the Brazilian states of Acre and Amazonas and represent different sources of resistance to the earlier hybrids. Hopefully, this resistance, along with a novel source of resistance identified in 2016 (Royaert et al. 2016) will be sufficiently distinct and durable to last.

So, how does resistance actually stop the witches' broom fungus? As we saw with resistance against black pod, the cacao tree has a formidable arsenal of defenses at its disposal—the trick is to get them activated. Cacao, as with all plants, is in an arms race with the various pathogens that are coevolving with it. This race is all about recognition, and at any point in time, the winner will be the one staying on top of the recognition game. Today, with witches' broom causing so many problems, it is clear that the fungus has successfully avoided recognition by the cocoa tree, meaning that the tree's defenses are not activated when the fungus attacks. But this will change, as plant breeders come up with new cacao varieties with resistance to witches' broom. Here, the cacao tree will be ahead in the recognition game and any attempted infection by the witches' broom fungus will elicit a defensive response.

The cocoa tree is rich in polyphenols, some of which, various flavonoids, for example, are used by the tree to defend itself against attack. One class of flavonoids, the procyanidins, are polymers of two compounds we've come across before—catechin and epicatechin, both of which are thought to have benefits for human health. But they are also good for the health of the cacao tree. Researchers at Rutgers University in New Jersey, USA, found that a resistant cacao genotype contained more procyanidins that a susceptible genotype, and when the resistant genotype was challenged by the witches' broom fungus, procyanidin levels increased further (Chaves and Gianfagna 2007). This is bad news for the witches' broom pathogen, since procyanidins are toxic, disrupting the germination of its spores and the subsequent development of the fungus.

Cacao also contains caffeine, which, in addition to acting as a stimulant, is important in the plant for its toxic effects against insects, fungi, and bacteria. Infection by the witches' broom fungus leads to an 800% increase in caffeine levels in cacao stems, which won't do the invader much good, since it can't grow in the presence of the stuff (Aneja and Gianfagna 2001). Young, actively growing leaves had little, if any, caffeine, but treatment with compounds that can trick plants into activating defenses led to caffeine accumulation. Trying the same trick with older, mature cacao leaves did not work, because caffeine did not accumulate. It seems, therefore, that young cacao leaves, which are much loved by the witches' broom fungus, can induce accumulation of this defensive compound when needed.

Tricking the plant into activating its defenses might be a useful way to tackle witches' broom in cacao varieties with poor or failing resistance to the pathogen, which, at present, seems to be most cacao varieties. This approach is known as induced resistance and works by circumventing the recognition that is required before plants activate their defenses (Walters et al. 2013). After all, the defenses the plant needs are there, so why not make use of them? It has already been tried experimentally on the cacao tree and shown to work. Using the commercially available inducing compound acibenzolar-S-methyl (ASM), Brazilian researchers showed that witches' broom could be reduced by 85%. As a bonus, the treatment also reduced the severity of another devastating cacao disease, verticillium wilt (Resende et al. 2002). The protective effect can be reasonably long-lasting, so applying ASM 30 days before challenging the young cacao plants with the pathogen provided good protection. This approach has received little attention in cacao, but it might be a useful addition to the tools required to tackle not just witches' broom, but other pathogens too.

Now, you might think there is nothing much the plant can do to defend itself against the witches' broom fungus until it tries to get into the plant, when it is likely to be met with fire and fury. But it seems that leaves of the cacao tree are like good Boy Scouts—they are well prepared. Cacao leaves are covered with short glandular trichomes—small hairs packed with proteins with antimicrobial properties. These chemicals are secreted onto the surface of the cacao leaf, where, along with proteins from within the leaf secreted through natural openings onto the leaf surface, they can really spoil things for the witches' broom fungus. A cacao genotype resistant to witches' broom was found to contain more than twice the number of these terrifying trichomes as a susceptible genotype. The Brazilian and French researchers who carried out this research found that if cacao seedlings were placed under sprinkler irrigation, some of the proteins were washed off, leading to greater witches' broom infection. The proteins could be kept on the leaf surface to do their job of ruining things for the pathogen if watering was done by drip irrigation directly onto the soil surface (Almeida et al. 2017).

Witches' broom has bequeathed a deadly legacy on those parts of the world unfortunate enough to harbor it. Given its track record in South America, substantial and significant losses can be expected if the pathogen spreads to Asia and Africa. This spread would be catastrophic in cacao-growing regions already afflicted by black pod, cacao swollen shoot disease (see chapter 8), and vascular streak dieback (see chapter 9). Keeping witches' broom out of Asia and Africa will require strict observance of quarantine regulations. Irrespective of the future spread of the pathogen, cacao production will depend on the continued development of disease-resistant genotypes, which need to be made available to growers.

7

Frosty Pod Rot

As the 1800s were edging toward the twentieth century, cacao plantations in the northern reaches of South America were suffering the effects of disease. In the northeast, witches' broom was making its destructive way through cacao crops, while on the opposite side of the continent, a new disease was causing problems for cacao farmers in western Ecuador (Evans et al. 2013). The diary of a plantation owner in 1895 records that "most of the pods become white whilst maturing on the trees" (Jorgensen 1970; Evans et al. 2013). When the Dutch mycologist Constant Johan Jacob van Hall visited western Ecuador 15 years later, he described a problem affecting cacao known by locals as *helada* or frost, which produced abnormal growth of the pods and beans (van Hall 1914; Evans 2012). Popular opinion at the time was that this affliction was caused by unfavorably low temperatures, perhaps reflecting the frosty-white appearance of the pods. Harry Evans reckons that van Hall either never saw the condition in the field, or he visited the country during the long, dry intercrop period, because, as an experienced mycologist, he would surely have recognized the affliction to be caused by a fungus, especially since masses of spores are produced on affected pods (Evans 2012).

The frosty affliction had been around for some time before the plantation owner's record of 1895 in Ecuador. Research into cacao cultivation in Colombia, published in 1953, found that the rapid expansion of the crop in the 1830s was checked when, in 1851, much of the cocoa crop was destroyed by

Figure 7.1. Symptoms of frosty pod rot on a cacao pod caused by the fungus *Moniliophthora roreri*. This photograph was taken in Peru. By permission of Martijn ten Hoopen.

what was described as "a virulent velvety fungus growth developing into an impalpable dust and attacking the fruit only" (Holliday 1953; Evans 2016, 65). What is truly remarkable about this interpretation is that the germ theory of disease was still being developed in the 1850s (Agrios 2005). But there appear to be even earlier records of frosty pod. According to Gabriel Cubillos, symptoms matching those of frosty pod were described in documents recounting the devastating effects of an affliction of cacao in the east of Colombia in 1817 (Cubillos 2017).

As the mysterious malady began to cause serious losses to their cacao crops, plantation owners in Ecuador decided enough was enough; in 1917, with the Great War continuing its bloody progress in Europe, they sought help. This came in the form of James Rorer, who made the trip from Trinidad to investigate the destructive affliction. With the help of mycologist Professor Ralph E. Smith at the University of California, the condition affecting the pods was found to be caused by a fungus, identified as belonging to the genus *Monilia*, close to the fungus responsible for serious diseases of temperate stone fruits, *M. fructicola*. It took another 15 years before the fungus was formally named as *Monilia roreri*, after James Rorer (Evans 2016, 64–68). Ensconced within the fungal genus *Monilia*, the disease became known as Monilia pod rot or Moniliasis. But the fungus had not been identified correctly, and it wasn't until the 1970s that the true identity of the fungal culprit was revealed. The genus *Monilia* belongs to a class of fungi known as Ascomycetes, which includes not just the scourge of stone fruits, *M. fructicola*, but also the powdery mildews, which affect a huge range of plants globally. Careful research, helped by the advent of powerful new molecular tools, revealed that the frosty pod fungus was not an Ascomycete, but a Basidiomycete—the group of mushroom-forming fungi to which the witches' broom fungus belongs. Researchers discovered that, incredibly, the frosty pod pathogen was closely related to the witches' broom fungus, *Moniliophthora perniciosa*, and the decision was made to rename it *Moniliophthora roreri* (Evans 2016, 64–68). The two fungi might be related, but as Money states in his excellent *The Triumph of the Fungi* (Money 2007, 81), the frosty pod fungus is a sort of handicapped version of the witches' broom fungus. So, whereas the witches' broom fungus, like all self-respecting members of the Basidiomycetes, produces mushrooms, the frosty pod fungus doesn't. In the witches' broom mushrooms, the gills on the underside of the cap produce structures called basidia, in which sexual reproduction occurs, giving rise to spores known as basidiospores. With the mushroom-less frosty pod fungus, no sexual fruiting body has ever been observed and the fungus seems to represent an unusual case of an asexual species within a mostly sexual class of fungi (Aime and Phillips-Mora 2005; Díaz-Valderrama and Aime 2016).

Now, I know what you're thinking—what could possibly make the frosty pod fungus ditch its basidiospores? Well, how about the formation of the Andes? I can hear the collective snorts of derision, but in all seriousness, the uplift of the Andes several million years ago divided the continent and created the Chocó refuge of northwest Ecuador and western Colombia. Some researchers have suggested that this allowed divergence of species of *Theobroma* on opposite sides of the mountains. In a similar way the predecessor *of M. roreri* and *M. perniciosa* was also allowed to diverge. It seems that the western side of the Andes was left with fewer host species for the fungus, and these were probably scattered throughout the coastal and submontane forests. With the enormous bulk of the Andes in its way, the predecessor of *M. roreri* faced two problems: how to disperse its spores effectively and how to increase their chances of survival? The basidiospores were no use, since they were short-lived and dispersed over a short range. There was considerable evolutionary pressure, therefore, to develop more robust spores capable of surviving long-distance dispersal. The suggestion is that this robustness was achieved by getting rid of the flimsy basidiospores and beefing up the new spores. This created a formidable pathogen, the frosty pod fungus, which overtook both black pod and witches' broom in importance in those countries to which it spread (Evans 2016, 87).

In fact, the frosty pod fungus *M. roreri* is endemic on wild species of *Theobroma* in northwestern Ecuador and western Colombia and became pathogenic on cacao in the latter region. It appears that when cacao plantations in Ecuador switched from the indigenous Nacional variety to the more productive Forastero variety in the late nineteenth century, the strain of the frosty pod rot fungus capable of causing disease in cacao was accidentally introduced from Colombia. The consequences for the Ecuadorian cacao industry were catastrophic—frosty pod devastated plantations and led to the gradual replacement of cacao with bananas. You can get some idea of the impact of frosty pod on cacao in Ecuador from James Rorer's account of one plantation where annual yield fell from 30 tonnes in 1916 to just 1.6 tonnes by 1919 (Rorer 1926; Evans 2016, 84). Over time, growing cacao became the province of smallholders. But with the disease responsible for losses of up to 43%, cacao cultivation does not make for an easy life (Aragundi 1974; Evans 2016, 84).

For many years frosty pod rot was considered of minor economic importance worldwide, primarily because it had remained confined to western Ecuador and Colombia. Nevertheless, it spread gradually, reaching western Venezuela in the 1940s, Panama in the 1950s, and Costa Rica toward the end of the 1970s (Evans 2016, 84–85). It was first detected in Costa Rica in 1978 and by 1983 cacao production had fallen by 72%, while export of cacao beans plummeted by 96%, damage so far-reaching that the industry has not recovered (Bailey et al. 2018). By 2005, it had reached Guatemala, Honduras, Belize, and Mexico. Researchers reckoned that natural spread of the fungus south and east from its stronghold in western Ecuador and Colombia was not possible because the Andes presented an impassable barrier, although they were worried that expanding agricultural development increased the likelihood of accidental introductions by man. Unfortunately, these concerns were justified, because by the mid-1980s, frosty pod rot was established in Eastern Ecuador, and having gained a foothold there, it spread south along the eastern slopes of the Andes, getting to Peru in the late 1980s. The effect on cacao production in Peru has been dramatic, with losses of 40%–50% leading to the abandonment of some plantations (Evans et al. 1998). Meanwhile, the spread of the fungus continues, with reports of its arrival in Bolivia and Jamaica in recent times. As I write this in July 2018, frosty pod rot has not reached Brazil, although cacao pathologists reckon it is only a matter of time before it spreads wherever cacao is grown in the Western Hemisphere. This is like a living nightmare, because wherever it invades, it quickly becomes the yield-limiting factor in cacao production (Bailey et al. 2018).

The Frosty Pod Rot Fungus: A Brief Biography

The white frosting on the surface of an infected cacao pod is neither frost nor icing sugar, but powderlike masses of spores, which are liberated from the pod surface by wind or tree movement and then distributed by air currents. Apparently, an infected pod can produce up to 7 billion spores that can survive on pods for nine months, but they last just a month on harvested pods left on the ground in the field (Bailey et al. 2018). When one of these spores finds itself on a cacao pod, it germinates and makes its way into the pod either directly, by forcing itself through the epidermis, or indirectly via a stomatal opening. The spore is, however, rather fussy, because not any old pod will do—it prefers young pods that are still expanding. This means that cacao pods are most susceptible during the first three months of development. Once inside the pod, the fungus starts its biotrophic phase and, like *M. perniciosa,* grows slowly between the cells for up to 90 days. Here again, the food supply is pretty limited in the spaces between the nutrient-packed cells of the pod, but there is enough to get by, providing one takes it slowly. Then, for reasons yet to be discovered, the fungus, which up to now has shown its gentler, more benign side, enters the necrotrophic phase. Dr. Jekyll becomes Mr. Hyde and the rot really sets in. Cells and tissues within the cocoa pod are killed quickly, and after a prolonged period in which most pods probably show little in the way of symptoms, chocolate-brown lesions appear, coalesce, and soon cover the entire pod surface. The fungus then emerges from the death and destruction within the pod, covering the pod surface in the frosty bloom of its billions of spores (Bailey et al. 2013). The mess left within the pod depends on the age of the pod when it was invaded by the frosty pod fungus. Complete destruction of the beans is likely if pods are attacked when just a couple of months old, whereas most beans can escape unscathed if pods are more than more than four months old when the fungus strikes (Evans 2016, 78).

Ever since I decided to study plant science at university back in the mid-1970s, I have been asked why I didn't study something useful. I should be used to it by now, but it saddens me to think that so many people still do not appreciate the importance of plants. I mention this because only a few years ago I was asked, by someone I thought ought to know better, why we should bother to study the biology of plant pathogens. I thought he was joking, but I was mistaken. Anyway, my reply was that if we are to control plant disease effectively, we must understand the biology of the pathogen. I'm not sure he was convinced.

But a thorough understanding of pathogen biology is the foundation for good disease control, in both plants and animals. With frosty pod rot, the source of inoculum is the infected pod, which, as we've already seen, produces billions of spores. These need wind and rain to deliver them to fresh pods and can travel up to 30 m on wind currents. Clouds of spores are dislodged from frosted pods by even the slightest breeze, or by raindrops. Incredibly, the spores can even be dispersed within a cacao farm by the convection currents generated by daily fluctuations in temperature. It seems there is no escaping these microscopic agents of destruction. But these malicious microbes like it wet—they need free moisture if the spores are to germinate and infect the pod. This moisture is provided by high humidity, and in Ecuador, for example, conditions are just right once the wet season starts in the December–January period. Huge numbers of spores continue to be produced on infected pods for several weeks, after which spore production plummets, although spores can still be collected from mummified pods more than a year after infection. It

Figure 7.2. Symptoms of frosty pod rot on three pods on a cacao tree in Peru. By permission of Harry Evans.

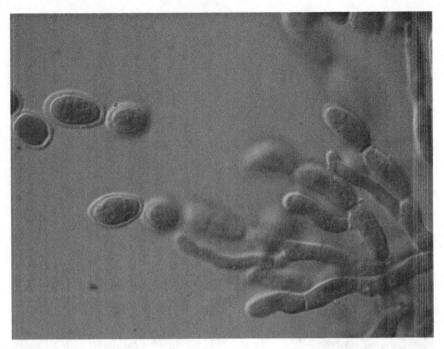

Figure 7.3. Spores of the frosty pod fungus *Moniliophthora roreri*. By permission of Harry Evans.

seems unlikely that 1-year-old spores would still be viable, but you never can tell, because 7-month-old spores have been shown to retain viability (Wood and Lass 1985, 296).

Stopping the Rot

Given the importance of infected pods as a source of inoculum, getting rid of frosted pods would seem like a good start to controlling the disease. In fact, regular harvesting and removal of infected pods before they start churning out spores by the billions has been shown to work in Ecuador, Colombia, Costa Rica, and Peru (Evans 2016, 88). Particularly important is the need to remove mummified pods before the rainy season starts, so the disease cycle is broken. A campaign using cultural control against frosty pod rot carried out in Colombia in 2005–2007 not only reduced disease incidence to minimal levels but was effective for three months (Cubillos 2017).

As with the black pod pathogen, humidity is important for spores of the frosty pod fungus, and any attempts to reduce humidity within the canopy of the cocoa crop would seem like a good idea. Optimizing shade limits humidity, and work in Mexico showed that reducing shade from 70% to 50% reduces levels of frosty pod (de la Cruz et al. 2011; Bailey et al. 2018). There is no doubt that good cultural control can help to limit the spread of the frosty pod rot fungus within the farm, but for these measures to be truly effective, all farmers must participate. Cultural control is labor-intensive and expensive, and if yields and market prices are low, farmers might be unwilling to adopt the measures. So, what about fungicides? In 1926, James Rorer reported that fungicides had to be applied at frequent intervals during active pod growth if they were to be protected from frosty pod rot, and he concluded that the cost was prohibitive (Rorer 1926). Not much has changed in the intervening 90+ years. Fungicides on their own are ineffective in controlling frosty pod rot, although used with other measures good levels of control can be obtained. But they are expensive, and so as with all the diseases we've looked at so far, breeding for resistance assumes great importance (Bailey et al. 2018).

In its center of origin in the submontane forests of the Chocó region of western Ecuador and Colombia, the frosty pod fungus evolved on endemic species of *Theobroma,* such as *T. gileri.* Growing on the pods of this relative of the cacao tree, the frosty pod fungus would have been parasitized by other microbes and fed on by insects. With this in mind, Harry Evans and colleagues scoured the northwest of Ecuador, looking for natural enemies of the frosty pod fungus. They discovered a guild of novel mycoparasites colonizing the fungus, as well as a number of bizarre invertebrates that feast on it (Evans et al. 2003; Evans 2016, 88–90). These invertebrates included the larva of a fungal gnat sporting a hooked abdomen, allowing it to cling to the pod while it munched on spores of the fungus. Attempts to identify the insect were thwarted because the pupae were themselves parasitized by a parasitic wasp, so they never made it through to the adult stage. It seems that this web of organisms feeding on the frosty pod fungus and which themselves are feasted on by others helps to keep the frosty pod pathogen in check in northwest Ecuador. So much so that up to 80% of *T. gileri* pods escape infection. By contrast, the frosty pod fungus does not appear to be significantly colonized and fed on by parasites and insects in any region where cacao is grown commercially. In fact, on cacao plantations in western Ecuador, if frosty pod is not managed, pod losses are often 100% (Evans 2016, 88–90).

Some of the fungi obtained from samples of *T. gileri* were found to be endophytic; that is, they are able to grow within the plant without causing harm and indeed, might benefit its host. One of these fungi, a species of *Trichoderma*, was able to colonize cacao seedlings without triggering the plant's defenses, which would put an end to its stay in the plant. This sounds promising, especially since some fungal endophytes can help the plant to fend off attackers (Evans et al. 2003).

Getting the better of the frosty pod rot fungus won't be easy, and recently published research provides a good example of why it is such a successful and devastating pathogen. The cacao tree, in common with other plants, possesses an innate immune system that recognizes and responds to potential attackers. It uses a surveillance system that recognizes fragments of the pathogen as not being of itself, whereupon it activates defenses against the attacker (Walters 2017). Chitin is a polymer that is a structural component of the cell walls of fungi, and fragments of chitin released in the early encounter between the plant and the fungal attacker are recognized as foreign by the plant, leading to the activation of defenses. Any pathogen worth its salt is not going to want lots of bits of itself floating around just waiting to be recognized by the plant. That would be akin to a burglar phoning up the police with his exact position just before breaking into a house. One way to avoid this is, somehow, to get rid of those bits of itself before they are recognized. The frosty pod fungus and its close relation, the witches' broom pathogen, have come up with an ingenious way to take the fragments of chitin out of action: they produce chitinases, enzymes that break down chitin, but which have been altered, rendering them catalytically inactive. In other words, they no longer break down chitin, but still retain the ability to bind it (Fiorin et al. 2018). In this way, any fragments of chitin released as the cacao tree and these pathogens begin their sparring are mopped up, reducing chances of the fungi being recognized by the cacao tree for what they really are—microbial thieves.

In an ideal world, cacao farmers would have access to varieties with good levels of resistance to frosty pod, and by using these and practicing good cultural control, the pathogen could be kept in check. So, in our less than ideal world, where do we stand in terms of resistance to frosty pod? Probably better than was originally thought. It has long been known that Forastero cacao is heavily attacked by the frosty pod fungus. This was noted by James Rorer as far back as 1918, who also observed that the local Ecuadorian variety Nacional was highly resistant to frosty pod rot. A large screening study carried out more recently revealed a cacao genotype from the germplasm collection in Trinidad, ICS-95, with significant levels of resistance to the pathogen. In fact, resistance to frosty pod rot seems more common than researchers thought, with scientists in Costa Rica identifying 278 clones exhibiting moderate or high levels of resistance to the pathogen. As far as researchers can tell, the frosty pod fungus reproduces clonally, meaning it does not involve sex. This might not seem of great importance, but it means that genetic diversity within the species is limited. We also know that resistance to the frosty pod pathogen is polygenically controlled (i.e., many genes are involved). Resistance controlled in this way is difficult for a pathogen to overcome, since it needs to tackle several genes instead of just one. There is hope, therefore, that frosty pod rot can be managed, and pod losses reduced (Bailey et al. 2018).

At present, Asia and Africa are free of frosty pod rot. Let us hope it stays that way, given its impact on the cacao crop and the difficulty of controlling it.

8

Swollen Shoots

Cacao cultivation in West Africa started in the eastern region of Ghana in the late 1870s when Tetteh Quarshie returned from Fernando Po with a few cacao pods. Before long, other farmers obtained pods from him and started growing cacao; soon its cultivation had spread throughout eastern Ghana. By the turn of the century, cacao cultivation had spread beyond Ghana to Côte d'Ivoire, Togo, and Nigeria. But, as we've come to expect, where cacao goes, diseases materialize, and true to form, in 1936 there was a report of trees developing swollen stems and dieback (Steven 1936; Wood and Lass 1985, 310).

This mysterious malady was probably affecting cacao trees as early as 1920 and perhaps as far back as 1907 (Posnette 1940), but its cause was unknown. The withering branches were thought to be due to decreased humidity resulting from clearing of the surrounding rainforest, thereby leading to water stress, while the stem swellings had been dismissed as unimportant (Posnette 2004). Whatever the cause, something had to be done, since the condition was killing cacao trees. Help came in 1937 in the shape of Adrian Frank Posnette (always known as Peter Posnette), who arrived in Ghana as a young plant pathologist with the Colonial Agricultural Service. Posnette hadn't long finished his postgraduate studies at the Imperial College of Tropical Agriculture in Trinidad when he traveled to Ghana. He then spent most of his first year on cacao farms, examining affected trees. His verdict was that the leaf symptoms and stem swellings were not the result of drought, but were caused by a virus, which was subsequently named cacao swollen shoot virus (CSSV).

Figure 8.1. Swelling of a chupon caused by cacao swollen shoot disease. By permission of Harry Evans.

By 1938, cacao production in the Eastern Province was being reduced to such an extent by diseases and pests that the Gold Coast Department of Agriculture decided to establish the Central Cocoa Research Station at Tafo to investigate the problem. The start of the Second World War a year later meant that only limited work could be undertaken, and by the end of the war in 1945, CSSV had killed so many cacao trees in the Eastern Province that cacao production was halved (Wood and Lass 1985, 311). In the meantime, workers found that cutting out trees infected by CSSV could contain the disease, provided there were frequent reinspections and removal of any newly infected trees. Wide-scale cutting out of diseased trees started in 1947, and by 1982, some 185 million trees had been destroyed. Even so, it was estimated that more than 30 million infected trees awaited removal, and unfortunately, the rate at which new infections were occurring far outstripped the rate at which they could be cut out. Twelve years later it was estimated that nearly 200 million cacao trees had been removed from 130,000 ha of land, while the latest estimates (2018) put the loss of trees at more than 300 million (Andres et al. 2018). Despite the enormous loss of cacao trees, the disease continues to spring up in new areas (Ampofo 1997; Muller 2016, 340).

Cacao swollen shoot disease spread widely and rapidly throughout the cacao-growing regions of West Africa, aided by the uniform planting of susceptible West African Amelonado cocoa. The disease reached Côte d'Ivoire in 1943, Nigeria in 1944, Togo in 1949, and Sierra Leone in 1963 (Muller 2016, 340). Once a tree becomes infected with CSSV, the prognosis is bleak. Severe strains of the virus can reduce yield of mature trees by 25% after the first year, 50% by the end of the second year, and after three years the tree will be dying, if not already dead (Brunt 1975; Ofori et al. 2015). Unfortunately, this devastating disease is endemic to West Africa but, thankfully, has never been reported in South America. Neither has it been reported in Sao Tome or Fernando Po, islands which were the main stepping stones for introducing cacao from South America to West Africa (Muller et al. 2018).

A Virus and Its Vector

As parasites go, viruses are pretty pared down. So pared down, in fact, that they comprise just a central core of nucleic acid (RNA or DNA), protected by a protein overcoat. Inside its snug coat, CSSV has a DNA core, and like all viruses, it uses the genetic machinery of its host to make more copies of itself. If you think this is the height of parasitic laziness, viruses even use other organisms to get to new host plants. In the case of CSSV, the transport of choice is the mealybug, as Peter Posnette discovered back in 1940. We now know that CSSV can be transmitted from one cacao tree to another by 14 different species of mealybug. Commonly known as scale insects, these sap-suckers appear as tiny, white, cottony masses on leaves and stems. Actually, the mealybugs are covered in a whitish "mealy" wax, which helps to reduce water loss from their soft bodies. They are found in warm, moist climates, but get introduced into greenhouses elsewhere, where they can be serious pests. Many of the mealybug species found on cacao are tended by ants; the ants feed on honeydew produced by the bugs and in return act as bug-bodyguards, protecting them against predators (Wood and Lass 1985, 314; Muller 2016, 345).

According to entomologists, mealybugs are sexually dimorphic. Now, before your imagination runs wild, all this means is that females and males are different. Females appear as nymphs and as adults they lack wings, whereas males are smaller and adults have wings. One common cacao mealybug, *Formicococcus njalensis,* lives for just six weeks, but in that time, it passes through three nymphal stages before becoming an adult. The first nymphal stage is more abundant than the others and is therefore more important in spreading CSSV than the other nymph stages. If the mealybugs find themselves on an infected tree, the young nymphs and adult females feed by inserting their needle-like mouthparts (a stylet) into the food-conducting tissues of the tissue (known as the phloem), where they take up not just sugars and other foodstuffs, but also the virus. Now full of CSSV, they take the deadly cargo with them wherever they go. For the later-stage nymphs and adult females this won't be very far, since their crawling will take them only to adjacent trees, although they can travel much greater distances if caught by the wind. The very active first-stage nymphs are also capable of spreading the virus over greater distances (Cornwell 1958; Thresh et al. 1988).

How Many Viruses Does It Take to Cause a Disease?

There is more to CSSV than swollen shoots. In some cacao varieties, veins in young leaves turn red, and then disappear. Depending on the variety and the strain of the virus, leaves can turn yellow along the veins, or flecks and spots can appear. Trees infected with virulent strains of the virus can produce small, rounded pods, which may develop a green mottling. Many strains of the virus also induce root swellings—a less than obvious symptom in mature cacao trees (Muller 2016, 342). If the cocoa variety is particularly susceptible, trees lose their leaves, branches and stems die, and eventually, so does the whole tree.

Until 2003, it was thought that cacao swollen shoot disease was caused by CSSV and two other cacao-infecting viruses—cacao swollen shoot CD virus (CSSCDV) and cacao swollen shoot Togo A virus (CSSTAV). Then in 2017, another virus was discovered in infected cacao trees in West Africa. This new virus, named cacao red vein virus (CRVV), was isolated from locations in Ghana and Côte d'Ivoire where "rapid decline and death" outbreaks had occurred (Chingandu et al. 2017). A year later, researchers discovered 7 further viruses associated with cacao swollen shoot disease, bringing the total number of viruses now associated with the disease to 10.

Containing the Virus

So, how do you control a disease caused by 10 different viruses? With great difficulty. Getting rid of infected plants is usually an effective means of reducing spread of a pathogen. So why, after what has been called the costliest eradication campaign of its kind anywhere in the world (Ampofo 1997; Thresh et al. 1988), did cutting out of diseased cacao trees in Ghana fail? One reason is that the eradication campaign was interrupted and discontinued several times. But another reason lies with the ineffectiveness of the eradication. Initially, the plan was to remove only trees showing symptoms of cacao swollen shoot disease. Adjoining trees within 30 paces were inspected frequently and treated as required. It might have been more effective if adjacent trees showing no symptoms were also cut out. But you can see the difficulty here—trying to convince farmers that they should destroy perfectly healthy-looking trees. Nevertheless, tree removals did continue, with farmers getting payments for trees lost or having trees replaced at their government's expense. Several innovative measures were also introduced, including "block planting" in the 1950s, in which large areas were cleared and replanted in contiguous blocks, and the "plant-as-you-cut" scheme of the 1970s, which involved the government removing trees, replanting new material, and then maintaining the farms for a period before handing them back to their owners (Andres et al. 2018).

Some of the problems facing the effectiveness of the eradication campaign were biological. For example, the insect vector can spread the virus from infected cacao trees even before they show symptoms, so by the time the trees start exhibiting symptoms, the adjoining, apparently healthy trees are likely to be infected. That is why, as we saw in the previous paragraph, farmers often paid the price for failure to remove seemingly healthy trees adjacent to those with symptoms. Another problem is some forest shade trees can also harbor viruses. The sheer size of these trees makes it difficult to spot symptoms of the disease, but there is also the issue of removing shade trees and the impact on the cacao trees below (Legg 1982; Wood and Lass 1985, 317).

But shade trees might actually be useful in the fight against cacao swollen shoot disease. Some studies have shown that shade trees can reduce pest populations and favor natural enemies of pests (Opoku et al. 2002; Thorlakson and Neufeldt 2012). Also, like most of us, the mealybug vectors of CSSV seem to like a bit of sun, since they are more abundant under full sun than shade. This might have something to do with the fact that their predators and parasitoids seem to prefer shade. Mealybugs aren't alone in liking a bit of sunbathing. Mistletoes also favor sunny conditions, because their seeds are spread by birds, which are most abundant at the forest edge. Now, you're probably wondering what mistletoes have to do with cacao swollen shoot disease. Well, the ants that tend the mealybugs are partial to mistletoe, and where the ants go, so do the mealybugs. This means that CSSV tends to be common where mistletoe populations are high—in unshaded cacao at the edge of the forest (Schroth et al. 2016).

Shade doesn't just affect the spread of CSSV, it also influences the severity of cacao swollen shoot disease. Research in Ghana showed that shade reduced symptom severity, which in turn helped to reduce yield losses. This work suggested that agroforestry systems with about 50% shade provide the optimal balance between severity of the disease and cacao yield (Andres et al. 2018).

In a newly established cacao farm, CSSV is most likely to take hold in the outer rows of trees closest to a source of the virus. Work in Nigeria suggested that new cacao plantings in areas where CSSV is endemic should be separated from existing cacao trees by at least 10 m. Although this was endorsed for adoption in Ghana, it never found favor with Ghanaian farmers, who were not keen on leaving relatively large areas of land unused (Domfeh et al. 2016). But what if the land, instead of lying bare, could be used to grow another crop? There is some merit here, especially if the crop in question is not a host for CSSV; that is, it is immune. Results from a seven-year study in Ghana demonstrated that citrus, for example, grown between new and old cacao plantings, acted as an effective barrier to the spread of the virus, protecting the newly planted trees from CSSV present in the older cacao. What's more, such crops could generate income that would compensate farmers for the income lost through not growing cacao (Domfeh et al. 2016).

Growing cacao varieties with high levels of resistance to CSSV would be the ideal solution. Nearly all cacao grown in West Africa until the 1950s was of the susceptible West African Amelonado variety. Field trials in Ghana in the 1950s showed that the spread of the virus in cacao derived from Upper Amazon and Amelonado parents was about half that in pure Amelonado cacao, while progeny from crosses between Upper Amazon parents and Trinitario cacao also showed reduced spread of CSSV. In later research, although some cacao hybrids were shown to withstand CSSV infection in long-term field studies, sufficient quantities of planting material were not produced (Muller 2016, 352). Recent research in Côte d'Ivoire found high levels of genetic diversity in several cacao genotypes from the Upper Amazon exhibiting good resistance to CSSV. This is important, since it provides a useful genetic resource for use in breeding programs (Guiraud et al. 2018). Hopefully, this will lead to new planting material for cacao farmers, although being a tree crop, this is likely to take some time.

When a cacao tree is infected with CSSV, symptoms can take some time to become apparent, and it can take several years for the afflicted tree to die. During this period, the disease can be spread inadvertently by farmers, for example by sharing infected plant material. This is thought to be the main reason for the rapid spread of CSSV throughout the cacao-growing countries of West Africa. Halting this spread is a real challenge, long recognized by the governments of the affected countries. In 2018, the governments of two of these countries, Côte d'Ivoire and Ghana, announced a joint effort aimed at combating the disease. In Ghana, where some 300,000 ha of cacao are infected, they plan to invest US$33 million in an effort to replace nearly 23,000 ha of infected cacao, while in Côte d'Ivoire, the government has pledged to spend US$19 million replacing more than 100,000 ha (Bisseleua 2019). In Ghana, the program involves compensation payments to farmers, helping them to sustain their livelihoods during the process of tree removal and replanting. Keeping farmers on board is essential and will require educating farmers on the effects of CSSV and the rationale behind the tree removal and replacement program (Ameyaw 2019).

Of course, you need to ensure that the new cacao material you are planting is actually free of the virus. This means being able to detect the virus in the planting material. Luckily, an early detection tool that can pick out CSSV DNA in a sample has been developed and plans are afoot to deploy it in Côte d'Ivoire (Bisseleua 2019).

Tree removal and replacement should not be seen in isolation, however, and ultimately, a combination of approaches, involving replanting, host resistance, barrier crops, and agroforestry/shading systems, offers the best hope of controlling CSSV. To be effective, these approaches need to be deployed in an integrated, coordinated manner (Ameyaw 2019).

CSSD is still spreading in West Africa, although the speed and effectiveness of its movement into new cacao-growing areas remains unclear. What is clear, however, is the need for better information on the epidemiology of the viruses and their vectors, especially the transmission of the viruses by endemic mealybug species, and the importance of alternative hosts in providing reservoirs of the pathogen.

9

Vascular Trouble

Papua New Guinea (PNG) occupies the eastern half of the island of New Guinea and includes the neighboring islands of New Britain and Bougainville. The western half of the island forms the Indonesian provinces of Papua and West Papua. Cacao was introduced to PNG around 1900 when Trinitario-type seed was brought to the country, most likely from Samoa (Wood and Lass 1985, 578). At the time the northern part of the country was under German rule, while the south of the country had been colonized by the British since 1884. Both parts ended up under Australian administration until its independence in 1975. PNG exported its first cacao beans in 1905, but production was small and did not increase much over the subsequent years, so that by 1940 only 200 tonnes were exported. Most of the crop was grown on the island of New Britain, and much of the cacao planting there was badly damaged during the war. The end of the war saw a rapid expansion in production, with 1,000 tonnes produced in 1950 increasing to 30,000 tonnes in 1970 (Wood and Lass 1985, 578). Production levels have stayed much the same since, and in 2017 output was 40,000 tonnes (Statista.com n.d.).

Two major problems continue to affect cacao in PNG: pod borers, which we will look at in the next chapter, and a fungal disease known as vascular-streak dieback (VSD). In the early 1960s, a serious dieback was affecting cacao trees in the Gazelle Peninsula on New Britain. Cacao was long known to suffer from dieback, with gradual loss of leaves and death of branch tips.

This dieback is caused by several factors, including lack of shade, nutrient stress, and damage from insects—ideal conditions for invasion by a range of pathogens just waiting for an easy way into the cacao tree. The dieback in New Britain was attributed to infection of weakened stems of the cacao tree by the fungus *Lasiodiplodia theobromae*, because this pathogen was regularly isolated from diseased samples (McMahon and Purwantara 2016, 310–311). But a young Australian PhD student, Philip Keane, working at the Lowlands Agricultural Experiment Station at Keravat on New Britain, noticed that the first symptoms of the disease occurred well before the branches invaded by *Lasiodiplodia* died. These initial symptoms occurred on leaves behind the branch tip, and, unlike the dieback caused by stresses, this version affected young healthy trees. Together with his PhD supervisors Ken Lamb and Noel Flentje, Keane found a fungus growing in the xylem or water-conducting vessels of affected trees on which the leaves had begun to turn yellow prior to falling off. The fungus was subsequently found to be a new pathogen of cacao, a basidiomycete eventually given the name *Ceratobasidium theobromae*. It produced its fruiting bodies during wet weather, with its spores appearing at night following an afternoon of rain. The fungus is a real night-lover, since the majority of spores are released between 1 and 3 a.m. Once liberated from their fruiting bodies, the spores are carried on the wind and are able to infect young cacao leaves. Once inside the leaf, the fungus colonizes the xylem vessels, which its hyphae use as a quick route to gain access to adjacent leaves. The VSD fungus is an aggressive pathogen, capable of killing cacao seedlings, causing severe dieback of mature branches and even death of the tree (Keane et al. 1972).

The decision was made to call the disease "vascular-streak dieback" to distinguish it from other forms of dieback that affect cacao. The name is apt, because one of the symptoms of the disease is a dark streaking of vascular tissue, in this case the xylem. When plants are infected by pathogens that colonize the xylem vessels, they respond by depositing toxic phenolic compounds. When an infected stem or branch is cut, the deposited polyphenols are oxidized, turning the vascular tissue brown. Plants also pump gels into the xylem vessels, while surrounding cells extrude their contents into the vessels, producing what are known as tyloses. The gels and tyloses restrict the movement of the pathogen within the vessels, helping to limit the spread of the pathogen to other parts of the plant.

Beginning in 2004, a new symptom was observed in plants infected with the VSD fungus. The margins and tips of infected leaves turn brown or black as necrotic lesions form. These lesions expand over several weeks, after which leaves fall off the tree. The precise cause of these new symptoms remains unknown, although it has been suggested that climate change might be involved. The widespread and rapid appearance of the new symptoms suggests a change in environmental factor(s) occurring on a regionwide scale. Possible candidates include increased levels of carbon dioxide or a rise in temperature. Another possibility is a decline in soil fertility—soil analyses from cacao farms in Sulawesi have revealed low pH and nitrogen content, and very low levels of organic matter (McMahon and Purwantara 2016, 324).

VSD was subsequently found in Malaysia, Indonesia, and elsewhere in Southeast Asia, but has not been found in other cacao-growing regions. This is just as well, since the disease nearly destroyed the cacao industry in PNG in the 1960s. Things only got better once cacao varieties susceptible to VSD were killed, leaving varieties with some resistance to the pathogen. These partially resistant varieties were propagated and grown, and eventually the industry returned to its former levels of production.

A severe VSD epidemic in Malaysia in the 1950s and 60s led to hybrid screening programs that eventually produced several clones with good resistance to the pathogen. These clones are now widely grown in Malaysia and Indonesia. Nevertheless, severe outbreaks of VSD have been experienced in West Java and Sulawesi more recently, leading farmers to replace cacao with other crops. The disease is so destructive that in their annual report for 2008/2009, the International Cocoa Organisation (ICCO) identified VSD as a major constraint to cacao production in the region (McMahon et al. 2018).

The cacao tree is the only known host for the VSD fungus, which suggests that it is a new-encounter disease, with the fungus transferring to cacao from a native host. It is possible that the fungus is not pathogenic on its indigenous host; if this is the case, without symptoms to give away its presence, finding the native host(s) will be difficult. Philip Keane writes that on finding heavy VSD infection on cacao planted near rainforest, he ventured into the forest searching for a possible source of the pathogen. He was overwhelmed by the sheer profusion of vegetation, with tall trees and their inaccessible canopies covered in lianas and epiphytes (Keane 2010). Many of these epiphytes are orchids, whose roots harbor symbiotic fungi, some of which are related to the VSD fungus. One of these fungal relatives might have taken a fancy to the newly arrived cacao tree, turning from friendly fungus to deadly pathogen in the process (Keane 2010).

Although most new infections arise from spread between cacao trees, some researchers reckon that the VSD fungus moves repeatedly from its native host to cacao, especially since cacao is often planted near rainforest (Keane 1981; Keane and Prior 1991). The prevalence of VSD in mainland PNG and eastern New Britain and its absence from the nearby New Ireland, Bougainville, and Solomon Islands, suggests that the native host harboring the fungus is also absent from these islands (Keane and Prior 1991). Keeping the fungus out of those parts of Southeast Asia that are currently free of it and preventing its spread to other cacao-growing parts of the world requires quarantine control. Although VSD is not transmitted via the beans, it could spread to other areas by importation of infected seedlings. Since symptoms of the disease don't become apparent for several months, the recommendation is that seedlings should be put into quarantine for six months (Prior 1985).

Being a systemic disease, VSD is difficult to control using fungicides or cultural methods. Pruning to get rid of infected branches provides some control if it is done regularly. Monthly pruning is more effective than carrying it out at three-monthly intervals, but this is likely to be much too labor-intensive for smallholder farmers, who tend to rely on family to carry out tasks on the farm (Prior 1980; McMahon and Purwantara 2016, 328). Ensuring that seedlings are protected is also important, since they do not survive VSD infection. Given that the fungal spores tend to be borne on the wind for distances of less than 100 m, it is recommended that seedlings are raised at least 150 m away from mature cacao trees in affected areas. Providing cover in seedling nurseries is also a good idea, because it prevents spores landing on leaves.

As we've already seen, cacao harbors a multitude of endophytes—fungi and bacteria that live quietly within the plant, causing no harm, and some of which provide benefits to the plant, including defense against pathogens and pests. So when a consortium of researchers in Indonesia and the USA examined cacao for endophytes, they found a fungus, *Trichoderma asperellum*, which turned out to be pretty useful (and which we've come across before—see chapter 5). The researchers found that spraying cacao seedlings with this fungal endophyte protected the plants against VSD (Rosmana et al. 2015). Subsequently, researchers working in Malaysia demonstrated the effectiveness of another species of *Trichoderma*, *T. harzianum*, in controlling VSD in mature cacao trees (Vanhove et al. 2016). Given the difficulty of controlling this very damaging pathogen, biological approaches such as this certainly merit further investigation.

Cacao trees with the ability to ward off the VSD fungus would be the ideal way to tackle the disease. We've already seen how effective even partial resistance is. Witness the survival of partially resistant genotypes following the epidemic in PNG in the 1960s, when susceptible types were wiped out. This resistance, though not complete, is obviously durable and, as such, is likely to be controlled by several genes; that is, it is polygenically controlled. Researchers are engaged in selecting cacao genotypes with resistance to VSD combined with good yield and quality characteristics. As with all research, the path is rarely smooth. Trials in Sulawesi, established in 2008 and evaluated between 2010 and 2012, identified cacao clones with promising yield and quality, but which were susceptible to VSD. Nevertheless, some clones exhibiting promising resistance, together with good yield and quality traits, were identified and could be used in breeding programs (McMahon et al. 2018).

Be Wary of Wilt

Wilting in plants is a sign that the shoots are not getting enough water. If we notice our houseplants wilting, giving them a drink of water will soon have them turgid once more. But sometimes watering won't do the trick, especially if the reason the water is not reaching the shoots is not lack of water in the soil, but inability of the water to reach the shoot. This might be because something nasty is destroying the roots, preventing them from taking up water from the soil, or it might be because the tubes responsible for transporting water from the roots to the leaves, xylem vessels, are blocked. As we saw with the VSD fungus earlier, some pathogens target the plant's vascular system, often with disastrous consequences, certainly for the plant. Some pathogens are vascular system specialists and none more so than those fungal pathogens known as "vascular wilt fungi." These include the fungi responsible for the devastating Dutch elm disease, *Ophiostoma ulmi* and its more aggressive relative *O. novo-ulmi*. Elm trees across North America and Europe were hammered by these xylem-loving fungi in two waves in the twentieth century, first by *O. ulmi* and then by *O. novo-ulmi* (Money 2007, 25–44).

A similar disease affects the cacao tree, causing wilting and death of branches and whole trees. Not surprisingly, it is caused by a fungus, *Ceratocystis cacaofunesta*, closely related to the Dutch elm disease fungi. This disease was first reported in Ecuador in 1918 but was not thought to be important, because the Nacional cacao variety grown at the time was resistant (Wood and Lass 1985, 330; Cabrera et al. 2016, 393). A more virulent form of the pathogen was detected in Ecuador in 1951, with Criollo cacao more severely affected than Trinitario and Forastero varieties (Wood and Lass 1985, 330). The pathogen has been reported in many countries in Central and South America, where it has been responsible for substantial mortality of cacao trees. Despite this, it is considered a serious "emerging" disease of cacao in Central and South America.

This fungus takes a laid-back approach to getting into the cacao tree—it enters passively through wounds. So, pruning with a contaminated machete is a surefire way of delivering the fungus to an easy entry point, hence the common name for the disease in Spanish: *mal del machete*. The fungus also gains easy access via wounds caused by bark beetles carrying its spores. Entry achieved, the fungus infects the cells surrounding the xylem vessels, eventually reaching the vessels themselves (Cabrera et al. 2016, 397). If the cacao genotype is resistant to the fungus, very little colonization occurs. If the cacao variety is susceptible, however, tissues are overrun with the fungus, which kills cells and tissues as it progresses and quickly goes on to produce spores. The plant responds by trying to block movement of the fungus in the xylem vessels using gums and tyloses, as we saw with VSD, and tries to kill it by depositing toxic phenolic compounds in its path. But with plant defense, timing is everything, and in a susceptible cacao tree, all these defenses come too late, with fungal progress too far advanced and aggressive to be stopped. Failure to stop the fungus means that its spores can be released into the xylem sap to be transported to other branches and leaves. In this way, large parts of the tree can be colonized by the fungus in quick time. Commonly, leaves start to wilt and turn yellow within two to four weeks of infection. Before you know it, it's curtains, first for the branch, and eventually for the tree.

For a pathogen that gains entry to its host via wounds, it is crucial to ensure that pruning tools are disinfected prior to use. Trying to control the fungus or the beetle chemically is expensive and not always effective and, as with other diseases of cacao, breeding for resistance offers the best hope for durable control. International programs aimed at breeding for resistance to *Ceratocystis* wilt started in the 1980s, resulting in the identification of cacao genotypes that were resistant under local conditions (Gardella et al. 1982; Cabrera et al. 2016, 412). Unfortunately, much of the cacao germplasm exhibiting good resistance to witches' broom is susceptible to *Ceratocystis* wilt (Engelbrecht et al. 2007).

But there is hope, because researchers in Brazil and the USA recently identified several candidate genes that will be useful in programs aimed at breeding resistant varieties (Fernandes et al. 2018).

So far, we've concentrated on diseases that pose a major threat to cacao. But diseases are only part of the threat faced by the cacao tree and those who grow it. As we will see in the next chapter, a range of insects also feed on cacao and in so doing cause damage, loss of yield, and even death of cacao trees.

10

Sap Suckers

Humans aren't the only animals with a taste for cacao. Insects also have a liking for it, so much so that more than 1,500 different types of the six-legged pests feed on the plant (Wood and Lass 1985, 366). In fairness though, only a tiny fraction of these insects cause damage of economic importance, and of these, most are of geographically local importance. In West Africa, for example, two species of sap-sucking mirid bugs cause major problems, while in Southeast Asia, the cocoa pod borer is the most serious invertebrate villain. Some insects damage cacao directly by feeding on leaves, branches, or pods, and to make matters worse, the wounds they inflict can be used by microbes just looking for an easy way into the plant. Other insects don't cause much damage themselves, but they carry a deadly cargo, notably viruses such as cacao swollen shoot virus (CSSV), which gain entry to the cells of cacao tree when they feed.

Mirids

Measuring in at less than 12 mm long, mirids are small. But what they lack in size, they make up for in looks, although beauty is in the eye of the beholder. They tend to be oval-shaped or elongate, and because they carry their head bent down, they have a hunched look. Think of an insect version of Scrooge from Charles Dickens's *A Christmas Carol*, and you'll have some idea of their general appearance. Still, some are brightly colored, and others have attractive

Figure 10.1. *Helopeltis* mirids on a cacao pod. By permission of Harry Evans.

patterns, although many are pretty drab. For the most part, they are what you might call "inconspicuous" (Wheeler 2001). Mind you, there are more than 10,000 species of mirid, and new species continue to be described.

These bugs are found on cacao in West Africa, where they are known as capsids, and in Asia and Central and South America, where they are called mirids. In Malaysia, they are commonly known as mosquito bugs or bee bugs. The two important mirid species in West Africa are *Sahlbergella singularis* and *Distantiella theobromae,* while *Helopeltis* species are important in Asia. In South and Central America, *Monalonion* species are found infesting cacao (Vos et al. 2003).

Mirids have been causing problems for cacao in West Africa for some 90 years and have been troublesome for even longer in South America. In some countries, it was several decades following the cultivation of cacao before the first reports of mirid damage started to trickle in. Take Papua New Guinea, for example—although cacao was grown there prior to 1905, the first mirid attack was reported in 1949, more than 40 years later. Cacao is obviously relatively new to many parts of the cacao-growing world, and mirids in these areas continue to adapt to the new plant (Wood and Lass 1985, 372).

A Mirid's Life and the Trouble It Causes

Mirids are most active during the warmest part of the day. They are good fly-ers, and once they alight on a cacao tree, they feed by pushing their needlelike mouthparts into the plant and sucking out the sap. This piercing and slurping results in small, water-soaked areas on the pods or stems that quickly turn black as the cells are killed by the bug's toxic spit. The two West African bugs tend to feed more on stems, causing severe and long-lasting damage to trees. When they feed on younger, soft stems, these can wilt and die. The wounds resulting from bug feeding allows opportunistic fungi to gain easy access into the cacao tree, leading to further damage (Wood and Lass 1985, 372–379).

The bugs lay their eggs on various parts of the cacao tree, including pods and pod stalks and the young vertical stems known as chupons. The eggs aren't just dumped on the plant, but are buried in the epidermal or skin layer, leav-

Figure 10.2. Symptoms of mirid damage on two cacao pods. This photograph was taken in Cam-eroon. By permission of Martijn ten Hoopen.

ing two slender filaments poking through the surface of the plant. These are breathing structures and can just be seen with the naked eye. The eggs hatch quickly, within one to three weeks, and then start a developmental process that sees them pass through five successive juvenile stages (nymphs). You will have noticed that one stage of development appears to be missing. That's correct, because mirids, in common with leaf hoppers, aphids, and scale insects, don't have a pupal stage. After about 30 days, a winged adult emerges, ready to start the process all over again. But their lives are short—they might live for a week, or if they're lucky and conditions are favorable, their lives might stretch to a month (Wood and Lass 1985, 372–379). Researchers have found that female *S. singularis* can live for more than 60 days, producing 236 nymphs in that time (Babin et al. 2011). Although females oviposited throughout their lifespan, the rate at which they did attained a maximum 3–4 days after the female reached sexual maturity and decreased gradually with age (Babin et al. 2011).

The damage inflicted by mirid attacks depends on the conditions under which the cacao is grown. Damage is widespread where the cacao is not shaded, whereas in shaded crops, damage is localized to unshaded patches. The latter can lead to groups of trees being severely affected, known as mirid pockets, while surrounding trees appear relatively unscathed. As you might expect, therefore, mirid pockets tend to occur in sunny areas of cacao plantations, where there are gaps in the canopy of shade trees. Studies in Cameroon showed that mirid pockets generally appeared in areas where amount of light transmitted through to the cacao trees was greatest (Babin et al. 2010). And yet, cacao trees grown without shade are unsuitable for development of mirid nymphs, which display negative phototropism during their development (Madge 1968). Subsequent work in Cameroon took things a step further by looking at the influence of tree structure on pests and diseases in a cacao agroforest (Gidoin et al. 2014). The researchers confirmed that the presence of forest shade trees was correlated with a decrease is mirid density, probably because large forest trees tend to even out the distribution of light throughout the canopy, limiting the development of mirid pockets. They also discovered that the reduction in mirid density was greatest when the forest shade trees were distributed randomly throughout the forest, rather than aggregated in clumps. The study also revealed that mirid density was affected to a greater extent by the availability of "sensitive" cacao tissues, such as leaf flushes and pods, than by the abundance of cacao trees in the agroforest (Gidoin et al. 2014).

Just how severe the damage is depends not only on the extent of mirid feeding, but also on subsequent wound invasion by opportunistic microbes, including several species of *Fusarium* and *Lasiodiplodia,* which cause dieback. There is also stress to the trees from lack of water and nutrients resulting from weed growth in unshaded patches (Wood and Lass 1985, 372–379). This interaction of factors accounts for damage that seems disproportionate to the number of mirids on the trees, with severely affected trees losing most, if not all, their leaves (Adu-Acheampong et al. 2014). According to some reports, an average daily population of just one mirid per tree can lead to the death of all cacao trees in an area (Nicol 1953; Adu-Acheampong et al. 2014). In fact, the economic threshold level for application of mirid control measures is just 0.6 individual mirids per tree in Ghana/West Africa and 0.7 individual mirids per tree in Cameroon/Central Africa (Decazy and Essono 1979; Padi and Owusu 1998). That scale of damage is staggering. Although damage is usually not quite that severe, losses are still high. A survey of Ghanaian farmers, published in 2016, revealed that cacao losses due to mirid attacks ranged between 30% and 40% (Awudzi et al. 2016). In Nigeria, yield losses of 30% to 70% have been reported, with losses of 75% on cacao farms where no control had been practiced for three years (Padi 1997; Anikwe et al. 2009). In Southern Cameroon, 68% of cherelles infested with *S. singularis* died (Mahob et al. 2019), backing earlier assertions that cacao fruits under three months old (i.e., cherelles) had little chance of surviving a mirid attack (Padi 1997). Clearly, doing nothing is not a good idea.

Controlling Mirids

Controlling any pest effectively requires an understanding of how its population is affected by seasonal variations in climate. Mirids are no exception. So, how do you determine the mirid population on a cacao farm? By counting them, of course! The method for doing this was described in 1971 and is known as the visual hand-height assessment method (Awudzi et al. 2017). Basically, the farmer inspects the usual mirid-inhabiting sites on each tree—pods, chupons, and flower cushions, for example—at hand-height, and makes a note of their numbers. This needs to be done between 6:30 a.m. and 9:00 a.m., when mirids are most likely to be active—like many of us, they don't like early mornings. As you might imagine, this is tedious, and for a farmer, very labor-intensive. But there is another way of obtaining mirid numbers—using pheromone trapping. Containers with mirid sex pheromones are placed in the cacao canopy and inspected every fortnight for trapped mirids, lured there by the prospect of sex, and their numbers recorded (Awudzi et al. 2017).

If you were expecting the two methods to give the same or even roughly similar results, you would be mistaken. With the hand-height counting method, mirid populations appear to peak between September and October, whereas with the pheromone trap, numbers peaked earlier in the year, during April and May. The problem is that the two methods don't provide the same information. The hand-height counting method picks up mostly nymphs and a few adults, while the pheromone trapping method attracts adult males. Clearly, using just one method will provide incomplete information. Nevertheless, what we know from this work by Awudzi and colleagues in Ghana is that populations of mirid nymphs start to increase in April, building up rapidly in June prior to the peak in September–October. Since nymph numbers start increasing in April, this would seem to be a good time to disrupt the population cycle. Applying insecticides at this point might be an effective means of preventing the buildup of adult mirid numbers (Awudzi et al. 2017). In Ghana at present, however, it is recommended that insecticides are applied every four weeks in August, September, October, and December. This advice is based on information from the 1950s, when mirid populations started to build up in July and peaked in October or November (Stapley and Hammond 1957; Awudzi et al. 2017). Perhaps there ought to be a reappraisal of the optimum timing for insecticide applications to control mirids in Ghana.

The organochlorine insecticide lindane was very effective against mirids in West Africa, but the bugs developed resistance to it, and it was superseded by various organophosphorus compounds. Lindane was eventually banned from use, and although some insecticides, including pyrethroids, provide good control of mirids, hazards associated with their use are prompting the search for less toxic alternatives. One approach has been to look to plants as a source of chemicals—biopesticides, if you like. Plants are masters of chemical defense, and some plants are stuffed with chemicals that can give an insect a really bad day. A plant bursting with insecticidal prowess is the neem tree, *Azadirachta indica* (Asogwa et al. 2010). The seeds of the neem tree produce an oil with serious insecticidal properties. These insecticidal effects are due to chemicals known as liminoids, which give neem a bitter taste. The most active of the liminoids in neem is azadirachtin (Asogwa et al. 2010).

Neem oil has received much attention as a possible biological-insecticide and in the past few years has been examined as a nontoxic, environmentally friendly means of controlling mirids on cacao. It acts as an antifeedant, putting insects right off what they thought was an easy meal, and to make matters worse, it can also inhibit the insects' growth. Cacao mirids don't last long when they ingest neem oil, with 98.5% dying within 96 hours of treatment with a 30% water extract of neem seeds (Asogwa et al. 2010).

When newly emerged adult mirids move from surrounding forest into a cacao farm, the first trees they colonize are semishaded. Once they mature, the adults move to areas within the cacao canopy with more light, and here, the females lay their eggs. Newly hatched larvae congregate in mirid pockets, where the tissues of the cacao tree face a double whammy of voracious larval appetites and high water loss due to exposure to sunlight. As we've seen, faced with these combined stresses, the cacao trees become water stressed and eventually die (Vos et al. 2003). One way to minimize this problem is to provide shade. In fact, one of the recommendations for controlling mirids is to maintain a uniform level of shade in cacao plantations (Babin et al. 2010). Care must be taken, however, since some shade trees, including various species of *Cola*, act as alternate hosts for cacao mirids. Not only that, but excessive shade can promote black pod. It seems therefore, that to be effective, shade management needs to provide a balance between conditions that are unfavorable to mirids and black pod (Babin et al. 2010).

Sometimes, natural enemies can be useful allies when tackling pests. A black ant going by the name of *Dolichoderus thoracicus* was used on cacao farms in Java to deter *Helopeltis* mirids as long ago as 1900. Although this practice seemed to have died out in the 1920s (Wood and Lass 1985, 382), the introduction of ants as part of a program of integrated pest management has been developed in Indonesia and Malaysia to control various species of *Helopeltis* mirids. First, however, antagonistic ants need to be removed using insecticide. You just wonder what else is being removed along with the antagonistic ants.

The predatory weaver ant, *Oecophylla longinoda*, seems to have a taste for mirids. In Ghana, an increase in weaver ant nests was associated with fewer pods damaged by mirids, as well as with cacao trees with the thickest canopies (van Wijngaarden et al. 2007). In this case, however, mirid control seemed to have more to do with the large ant population taking up so much space they simply squeezed out the mirids, rather than predatory behavior by the ants. Whatever the mechanism, mirid control with these ants is thought to be as efficient as using an insecticide, with the advantage of being less expensive (Offenberg 2015).

But ants aren't the only natural enemies that could be used to control mirids. Eggs and nymphs of cacao mirids are preyed upon by various parasitic wasps, and because they are very good at parasitizing the bugs, they are promising candidates for a more biologically based approach to mirid control (Vos et al. 2003).

Providing farmers with cacao varieties able to resist mirids is an important component of any integrated management program. Some cacao hybrids are less vulnerable to attack from mirids than others, and research in Nigeria, for example, has identified several cacao genotypes with good levels of resistance to the brown cacao mirid, *S. singularis* (Anikwe et al. 2009). In Côte d'Ivoire, researchers studying resistance to *S. singularis* and *Distantiella theobromae* found that Forastero genotypes from the upper Amazon appeared to be better parents for breeding than Trinitario clones or Forastero types from the lower Amazon (Sounigo et al. 2003).

One aspect of resistance to insects that can be overlooked is attractiveness of the plant. This is a shame, since making cacao trees less attractive to mirids could be a useful way of reducing damage. Research in Cameroon discovered that cacao genotypes differ markedly in their attractiveness to mirids, with some genotypes, including the Catongo Trinitario genotypes and one from the Upper Amazon (IMC60), being least attractive (Dibog et al. 2008). As it turned out, the latter clone was also the most tolerant of mirid infestation in the field. It appears that the lack of attractiveness to mirids is due to the low water content in their stems (Vos et al. 2003).

Sucking the life out of your host like an invertebrate, vegetarian dementor is bad enough, but the cacao tree must also contend with other equally unwelcome and damaging attackers, as we will see in the next chapter.

11

Cacao Borers

The minute moth, *Conopomorpha cramerella,* is just 7 mm long, and even its wingspan is only 12 mm. With bright yellow patches at the tips of its forewings and its very long antennae swept back, it's a dainty wee thing. It flies feebly and, in the air, looks rather like a slow-flying mosquito (Vos et al. 2003). But appearances can be deceiving, for this fragile-looking moth gives rise to cacao-loving offspring that bore into pods, with serious consequences. The larvae of this tiny moth drill through the husk of the pod, making long frass-filled galleries in the pulp. They don't eat the beans, but that is no consolation, because larval activity leads to callus formation, which screws up bean development. Pod contents turn into a hard, gall-like mass, in which beans are small, flat, and stuck together. Beans from such pods are unusable, and if the infestation is heavy, more than half the crop can be lost (Wood and Lass 1985, 403; Vos et al. 2003).

Cocoa pod borer is the most damaging insect pest in Southeast Asia, where it is thought to be endemic. It was first reported attacking cacao in Sulawesi in the 1860s and since then, it's been spreading across the Pacific. It reached the East New Britain province of PNG in 2006 and Bougainville Island two years later. An extensive eradication program undertaken in East New Britain was thought to have been successful, but subsequent monitoring showed that the pod borer had spread; the eradication program had failed. In this region, pod losses due to pod borer infestation can reach staggering proportions (Yen et al. 2010). The inexorable spread of the pest continues. It has recently been reported from Northern Queensland in Australia, where attempts are being made to prevent it becoming established (Valenzuela et al. 2014). Let's hope it works, because pod borer is responsible for average losses in cacao production of between 40% and 60%, with losses reaching as high as 90% in unmanaged farms. Such losses are estimated to cost a staggering US$500 million annually to the Indonesian cacao industry (ICCO 2015).

The Brief Life of the Cocoa Pod Borer

The delicate adult moth rarely lives for more than a week, but that is enough time for the female to produce up to 100 flattened, orange-brown eggs. The adults are lovers of the night, and this is when they mate, and also when the eggs are laid. The female lays them singly on the pod surface and seems to prefer placing them in pod furrows. About a week later, they hatch, and a miniscule (1 mm) larva—the first instar—emerges from each egg. The hungry larva bores it way through the pod husk, and once inside the pod, it starts to satisfy its appetite by feeding randomly. Although the cacao beans are not usually consumed, in young pods the larvae can penetrate them, leading to undersized and misshapen beans. After 15 days or thereabouts, the larvae are fully grown, having increased in length some 10 times. The ever-hungry, off-white larvae eat their way out of the pod through the husk and emerge pale green. On reaching the outside of the pod, the larvae find a suitable site to pupate, either by crawling to a suitable site on the pod surface, or by rappelling their way down on a silk thread, like a larval James Bond, to a leaf on the tree or a dead leaf on the ground. The multitasking larva then makes a waterproof silken cocoon and hunkers down for about a week, after which the adult moth emerges for the final week of its circle of life. And that's it—egg to moth in one month (Wood and Lass 1985, 402; Vos et al. 2003).

The pod borer certainly likes cacao, but it is also partial to rambutan (*Nephelium lappaceum*), nam-nam (*Cynometra cauliflora),* and *Cola nitida.* The last of these is an introduction to the region from Africa, and it is likely, therefore, that rambutan and nam-nam are its indigenous hosts (Wood and Lass 1985, 403).

Putting an End to the Boring Pest

In pest control, as in life, timing is everything. Get the timing right, and you can achieve good control; get it wrong, and the consequences can be disastrous. So, when is the best time to tackle cocoa pod borer? Obviously, in an ideal world, you would be able to prevent *any* pod borers entering the crop, but that, though laudable, is unrealistic. So, given that there will be pod borers around, it helps to have some idea of the level of infestation the trees can tolerate before yield starts to be badly affected. Research conducted in the 1980s revealed that pod infestations of up to 60% had little effect on yield. Losses increased rapidly, however, if infestations were greater than that (Day 1989; Vos et al. 2003). The level of infestation was determined by opening pods and assessing the extent of internal damage. In more recent research on the relationship between pod damage and yield loss, pods were grouped into one of four categories: A = no damage; B = light damage; C = moderate damage; D = heavy damage. Developed by researchers at Mars Inc., this is conveniently known as the A, B, C, D method of classifying pods for damage (Valenzuela et al. 2014).

So, armed with this knowledge, what comes next? You could try chemical control, but blanket spraying of insecticides in the hope of killing pod borers will also kill other insects, including many natural predators of insects that are usually minor pests of cacao. This approach is akin to shooting yourself in both feet—not advisable. Chemical control of pod borers is difficult, simply because the larva spends most of its life within the cacao pod. But, apparently, a contact insecticide such as a pyrethroid, applied to the undersides of the lower branches in the period between the main harvests, can keep pod borer levels below the threshold for economic damage (Vos et al. 2003).

Of course, you don't have to wait until pods are half destroyed before attempting to control pod borer. In the early 1900s, complete harvesting of cacao pods on a regular basis, known as "rampassen," was thought to be the only means of tackling pod borer (Yen et al. 2010). Later studies found that removal of all pods more than 6–7 cm long for six weeks disrupts the insect's life cycle because females prefer to lay their eggs on older pods (Vos et al. 2003). Harvesting pods when they are only just beginning to ripen ensures that most of the larvae are still within the pod. But the little pests need to be dispatched—there is no point in harvesting the pods and leaving them for the larvae to bore their way back out. Opening the pods and burying them or covering them with plastic, or even just putting the unopened pods into plastic bags for a few days, will kill the larvae by overheating. The main drawback to this approach is the continued movement of female moths from surrounding cacao farms where no such control is being practiced.

Preventing the moths from laying eggs on pods sounds like the ideal means of dealing with the problem. This has been tried, by enclosing young pods in thin plastic bags with open bottoms to allow adequate ventilation. The pods are then left in their protective sleeves while they mature. This method can provide complete control of pod borer, providing the bags are not put on the pods too late. As you can imagine, however, this method is very labor-intensive and therefore costly (Wood and Lass 1985, 402–407).

Sexual attraction is immensely powerful, especially when the main drive in life is to get your genes into the next generation. But this can be a fatal attraction, providing a potential means of managing insects. Imagine being lured by the irresistible scent of a receptive female only to find an empty trap. This is the idea behind using pheromones to control insect pests. It's a great idea, providing the males can be trapped before they mate. With pod borers, many more male moths could be trapped using synthetic pheromones than using female moths, and in trials, pod losses were reduced substantially. Things looked promising until a new, much fussier, race of pod borer was found that wasn't attracted to the pheromone mixture being used. Although a different pheromone mixture was developed and found to attract both races of the pod borer, the worry was that new races of the insect would emerge, requiring further additions to the pheromone cocktail. Work published in 2008 suggested that with improved quality of commercial pheromone preparations and better trap design, pheromone trapping could be used in monitoring, and maybe even in mass-trapping of pod borers (Zhang et al. 2008).

Here again, the development of cacao clones with resistance to cacao pod borer would be welcomed. Female moths prefer to lay their eggs on pods with well-furrowed surfaces, presumably because the eggs are less exposed and less likely to be washed off by heavy rains (Azhar 1990). Once the larvae have started to bore their way into the pod, the going can get tough, and pods of cacao varieties with thicker or harder layers in the pod wall can cause many larvae to give up the fight and perish. Not surprisingly, larvae prefer pods with thinner and/or softer walls, and as a result, more of them survive the journey into the pod. Cacao varieties with pods exhibiting suitable characteristics would be valuable in attempts to control pod borer, but breeding takes time, and there is also the time, effort, and cost involved in replacing trees with newer varieties. According to the International Cocoa Organisation (ICCO), research efforts are concentrated on developing genotypes with harder pod walls, since there is a high mortality of larvae in genotypes with either thicker or harder pod walls. Researchers are also screening selections for resistant material from the diversity of genotypes currently planted in Sulawesi (ICCO 2015).

Pod borer outbreaks can be devastating, and trying to prevent such outbreaks requires considerable effort and resources. For smallholders in certain parts of Southeast Asia, this might be too much, especially if they are trying to pick themselves up following a severe pod borer outbreak. A study conducted in the East New Britain province of PNG concluded that the changes in lifestyle, coupled with the intensive methods required to tackle pod borer effectively, were too much for many farmers, who opted-out of growing cacao (Curry et al. 2015).

Boring Gets Tough

If you thought boring into a cacao pod was impressive, just imagine boring into the stem of a cacao tree. But that is exactly what a triumvirate of enterprising insects do—they are stem borers. The bothersome bunch—two lepidopterans (*Eulophonotus myrmeleon* and *Zeuzera coffeae*), and a weevil (various *Pantorhytes* species)—are considered primary pests of cacao. *E. myrmeleon* occurs throughout West Africa, while the other stem borers are found in Southeast Asia and PNG (Vos et al. 2003).

The larvae of *E. myrmeleon* bore tunnels into stems of seedlings and mature trees that run lengthwise within the stem for up to 30 cm. You can tell when an active larva is in residence because a sticky sap dribbles down the stem from the entrance hole. When the gooey liquid dries, it leaves a dark water stain on the bark. In dry conditions, larvae will leave tunnels as the tunnels dry out and start new tunnels in moist wood. If dry weather persists, many of these tunnels will dry out, destroying stems in the process. If dry weather coincides with a heavy stem borer infestation, damage to cacao trees can be great. Even if conditions aren't dry, the entrance holes bored by the larvae allow microbes such as the black pod pathogen to enter, resulting in further damage to the stem, which can lead to wilting and death of the tree (Vos et al. 2003). The larvae pupate and after about three weeks, adults emerge. These handicapped grown-ups have no mouthparts and so find themselves on an unavoidable fast. Not surprisingly, their lifespan is brief—just four days. They act quickly and mate within a day of emerging from the pupa, and within an hour of mating, the female lays 500 eggs. It's amazing what can be crammed into a short life.

Larvae of *Z. coffeae* experience a risky aerial adventure before they begin the business of boring into the stem of a cacao tree. The adults, known as leopard moths, lay pale yellow eggs on stems and branches. After hatching, the gregarious larvae spin a communal web. Then, ninja-like, they lower themselves by silk threads, and with each thread acting like a parachute, they are carried away on the wind. Many perish on this wind-borne flight, but those fortunate enough to land on a cacao stem or branch bore their way into the bark (Wood and Lass 1985, 401–402; Vos et al. 2003).

The larvae of *Pantorhytes* don't do anything so adventurous, but the adult weevils, although unable to fly, can live for up to 15 months. This is pretty long-lived by stem borer standards, but those reared in the lab can live to the ripe old age of two years. Their larval stage is extended, taking between five and nine months to complete development. They bore their way about 2.5 cm into the stem, forming tunnels that run parallel to the stem surface. Several larvae drilling holes into a young stem can cause it to crack, which, ultimately, can be fatal to the stem (Wood and Lass 1985, 419–420; Vos et al. 2003).

Predators to the Rescue?

Controlling stem borer infestations is difficult, but one promising approach in PNG and the Solomon Islands is the use of ants to reduce numbers of *Pantorhytes* larvae (Vos et al. 2003). Here, cacao trees foraged by ants are less likely to have live larvae, so encouraging ant colonization might help to keep larval numbers at low levels. In practice, however, introducing the ants to plantations and getting them to colonize trees has been challenging.

But ants don't necessarily have to be introduced. They are an important part of tropical agroecosystems, in which they play various roles, including acting as pest controllers (Perfecto and Castiñeiras 1998). In Malaysia and Indonesia, for example, ants provide pest control in cacao farms by deterring mirids and the cocoa pod borer (Way and Khoo 1991). But ants also help other insects, by providing protection from predation, for example. When the insects they assist are important pests of cacao, such as mealybugs, their beneficial effects in providing pest control can be overshadowed.

Samantha Forbes and Tobin Northfield of James Cook University in Cairns, Australia, studied the effects of *Oecophylla smaragdina* ants on insect pests in cacao, publishing their results in 2017 (Forbes and Northfield 2017a). They looked at three important pests of cacao in Australia—*Helopeltis* mirids, fruit-spotting bugs (*Amblypelta* spp.), and black swarming leaf beetles (*Rhyparida nitida*)—as well as mealybugs, which although not important in Australian cacao farms, can carry cacao swollen shoot virus (CSSV). The ants did a good job of providing pest control, reducing numbers of the mirids, fruit-spotting bugs, and leaf beetles. Ants were observed capturing the pests and taking them back to the nest, suggesting that the ants were preying on the three insects. But another mechanism might also have been at play here, and it has to do with olfaction. *O. smaragdina* ants produce a chemical odor trail that provides directions to fellow ants—a sort of chemical recruitment aid, if you like. This chemical trail can last for several days and, in addition to guiding their colleagues, might also alert other predators to the presence of prey, and deter herbivorous insects from going anywhere near the odor for fear of being eaten (Forbes and Northfield 2017a). But this pest control came with a potential downside, because the presence of the ants increased mealybug numbers. This was not a problem in their study, since CSSV is not present in Australia, but it could be disastrous in areas with CSSV.

Managing the habitat on a cacao farm to promote natural pest control, such as that provided by ants, can be very effective. But natural pest control needs to be compatible with practices employed to promote pollinators in the cacao crop. Decomposing husks of cacao pods are used as a mulch on cacao farms to support populations of insect pollinators, possibly by providing them with an alternative habitat as well as food resources. In their study, Forbes and Northfield found that the natural pest control provided by the ants was compatible with the support for pollinators provided by using cacao pod husks. Indeed, another study by Forbes (2015) found that using cacao pod husks as a mulch beneath trees increased numbers of other types of predators, namely spiders and skinks. So, the cacao pests were faced, not just with ants, but with other predators too.

In cacao farms, as in other agroforestry systems, and indeed in natural forests, arthropods are predated by bats and birds. The diversity of these aerial vertebrate predators is likely to be different in a natural forest compared with a cacao farm. You might also expect differences in bird and bat diversity in cacao farms depending on management practices employed, such as shading. A study carried out in Brazil and published in 2016 found that excluding bats and birds from a cacao plantation affected arthropod numbers and damage to the trees (Cassano et al. 2016). Excluding birds led to increased damage to leaves of cacao trees, while excluding bats increased the abundance of arthropods and plant-sucking insects. In this study, shading exerted little influence. Earlier work showed that excluding bats and birds from a cacao farm decreased yields by 31% (Maas et al. 2013), while beneficial effects of bats and birds on pest control have also been reported for other crops, including coffee, cotton, and rice (Cleveland et al. 2006; Karp et al. 2013; Wanger et al. 2015).

Although many insects are capable of moving considerable distances, bats and birds are highly mobile and can travel far and wide. It is likely, therefore, that any management practices in support of wildlife would need to be applied at a considerably larger scale than an individual cacao farm.

Like an iceberg, a substantial part of the cacao tree, as with most plants, cannot be seen. I refer, of course, to the roots, which, although well hidden in the soil, are no less prone to attack, as we shall see next.

12

Going Underground

In my final year of undergraduate study, toward the end of 1977, I was looking for PhD opportunities and, after traveling to many universities and research institutes, decided to accept the offer of a PhD place at Lancaster University. I had taken an instant liking to Lancaster, the university campus, my future PhD supervisor, and the project. I was going to spend three years looking at how roots of barley were affected by the leaf-infecting powdery mildew fungus. I was thrilled and excited. So, imagine my surprise when I was told that I was wasting my time, when all the interesting stuff was happening aboveground, in the leaves. I knew this was nonsense and so I ignored it. I'm glad I did, because I ended up spending several wonderful years at Lancaster, working on roots.

There is a saying, that "out of sight is out of mind," and I think that even today, more than 40 years since my undergraduate days, things that happen belowground still tend to be overlooked. The tide has been turning over the past few years, however, revealing the complexity and importance of underground processes that were hitherto not given the attention they deserved, in part because they were out of sight. The importance of soil, its inhabitants, and everything that goes on in it is vital, not just to plants, which grow in it, but also, as a consequence, to the rest of life on this planet.

A huge variety of microbes and invertebrates live in the soil, and although the vast majority don't harm plants, some do feed on roots, causing damage and crop loss. A number of soil-dwelling insect species are associated with tropical tree crops, among them, a group most people would not think of as crop pests. I refer to termites, soil-dwelling insects which have increased in importance as pests of cacao over the years.

A Brief Look at Termite Biology

To some people, myself included, the mention of termites conjures up images of the huge aboveground structures they construct. Termite mounds, as these structures are known, are among the largest structures built by a nonhuman animal. Some of these mounds are centuries old and can reach truly massive proportions. They can be up to 9 m in height, which, if you consider the size of a termite, would be the equivalent of us constructing a building several times taller than the Empire State building. And it's not just the height that is impressive—so too are the interiors of the mounds. Termites use huge amounts of mud and water to build mounds with the most elaborate internal architecture—tunnels, galleries, archways, and even spiral staircases—not to mention the external architecture that can include chimneys, pinnacles, and ridges. What a home! Except the termites don't use these architectural gems as houses, they use them as a sort of gas exchange system, helping them to breathe. Like us, termites breathe in oxygen and exhale carbon dioxide, and the combination of the intricate internal design of the mound and external gusts of wind ensures effective exchange of these gases with the air outside the mound and also allows temperature to be controlled within the mound (Gullan and Cranston 2014, 344; Srinivasan 2018). The termites actually live several feet below the mound, in a nest with numerous gallery chambers. Mound architecture is diverse, reflecting the construction skills of different termite species. The "magnetic mounds" built by the debris-feeding termite of northern Australia, *Amitermes meridionalis,* have a narrow north-south and broad east-west orientation, which helps with thermoregulation. Here, the broad side of the mound gets the best of the early morning and late afternoon sun, while the narrow side is exposed to the high, hot midday sun (Gullan and Cranston 2014, 344).

But not all termites construct huge mounds—some make nests in wood, such as rotting timber, or even a living tree (Gullan and Cranston 2014, 508–509). In fact, termites display four types of nesting behavior: those that nest in and eat a single piece of wood, or several closely associated bits of wood; those that start in one piece of wood before moving on to another, moving their nest in the process; those that don't like to eat in bed, meaning they nest and feed separately; and those with no fixed abode, which includes soil-feeding termites that eat dead plants (Gullan and Cranston 2014, 344).

Termites are detritivores, meaning they eat mostly dead plants, although some are partial to living plants, including grasses and trees; some species even feed on that most unpromising of foodstuffs—wood. Many termite species have special organisms in their gut that break down the cellulose in their diet into sugars they can digest. Some species have protozoa in their guts, while others have bacterial inhabitants (Gullan and Cranston 2014, 344). Some termites also "tend" or culture fungi in combs of their own feces in the nest, and the mixture of fungi and feces is used to feed the colony (Gullan and Cranston 2014, 244).

Termites have a caste system, with workers at the bottom and the queen and her consort at the top. As their name implies, workers do most of the work within a termite colony—foraging, storing food, and maintaining the nest. They are unspecialized and weakly pigmented, the latter characteristic giving rise to their popular name—white ants. Defending the nest is the job of the soldier caste, which have anatomical and behavioral adaptations that help them in this role. Many soldiers have huge heads and large, powerful jaws. They also have a strongly pronounced snout, through which they squirt defensive chemicals (Gullan and Cranston 2014, 341). But these fearsome modifications come at a cost because the oversized jaws cannot be used for feeding. Instead, the soldiers need to be fed by the workers. These two termites, the nonreproductive castes, are wingless and in most species, completely blind. Then there is the reproductive caste—the queen and king, which mate for life. The queen is responsible for egg production for the colony, and in order to increase fecundity, her abdomen swells up enormously (Bignell et al. 2010).

Unlike ants, bees, and butterflies, which undergo complete metamorphosis (egg, larva, pupa, adult), the process in termites in incomplete. Here, the early stages of development look like smaller versions of the adult (Gullan and Cranston 2014, 341). These preadult stages, or nymphs, molt first into workers, with some workers molting into soldiers, and others into winged forms, alates. The winged forms leave the colony, with males and females pairing up before landing to find a suitable spot to start a colony. Only then do they mate, following which they never leave the nest. This can be as long as half a century, since a termite queen can live for anywhere between 30 and 50 years. The first offspring of the new royal couple are workers, which are fed regurgitated plant material containing the symbiotic gut microbes they need to enable them to digest the cellulose in their diet. When they are old enough to feed themselves, the workers start to enlarge the nest, making room for the expanding colony (Gullan and Cranston 2014, 344).

From Soil Engineer to Cacao Pest

In tropical regions, termites are probably the most important engineers of the soil ecosystem (Jouquet et al. 2011). They can forage over large distances (tens of meters), and their climate-controlled living quarters allow them to remain active in harsh environments and throughout the seasons. This means that in some habitats, such as arid and semiarid savannas, they are often the only invertebrate detritivores still active in the dry season (Jouquet et al. 2011). There can also be a lot of them. In the tropics, there can be as many as 7,000 individuals per meter, with some nests containing millions of termites.

As we've seen, many termite species feed on dead plant material, chopping it up with their mouthparts and grinding it in their gizzard, thereby increasing its accessibility to soil microbes (Jouquet et al. 2011). As a result of termite activity, organic matter is returned to the soil, via their feces, their bodies, and the nesting structures they build, material that would otherwise be lost in the fires that periodically sweep across the dry savannas (Jouquet et al. 2011). But they are not just decomposers of organic matter, they also loosen the soil, recycle nutrients, modulate the availability of resources for other soil-dwelling organisms, and create habitats for large number of animals (Jouquet et al. 2006).

Those termites with a liking for wood—that is, those that nest in and eat wood, about 80 species—can cause significant damage to property. The subterranean termite *Coptotermes formosanus* lives in underground nests, with colonies containing up to 8 million individuals. This is a termite with a serious record of damage to human constructions. Not only is it responsible for the undermining of dams in its native China, but it is thought to be partly responsible for the failure of the flood walls in New Orleans following Hurricane Katrina in 2005 (Gullan and Cranston 2014, 344). An additional 100 or so species are considered pests or a general nuisance, and it has been estimated that the cost of termite damage and control globally is around US$40 billion every year (Gullan and Cranston 2014, 352).

Termites have long been associated with cacao, and although they were once considered minor pests (Wood and Lass 1985, 424), their status has changed over the years. Today, termites are regarded as major pests of cacao (Ambele et al. 2018). Of the approximately 3,106 species of termites, 371 are considered pests, because in some ecosystems, their beneficial and important role in decomposition and nutrient recycling is outweighed by the damage they cause (Ambele et al. 2018). Termites affecting cacao include those belonging to the genera *Macrotermes*, *Nasutitermes*, *Microcerotermes*, *Ancistrotermes*, and *Coptotermes* (Ambele et al. 2018). Some of these termites are mound builders, such as *Macrotermes* spp., while others, *Nasutitermes* for example, construct carton nests of feces and wood on tree trunks, or small dome-shaped mounds on trees.

Twenty years ago, the termite *Macrotermes bellicosus* was not thought to damage cacao, but things change, and today, it is considered a pest of the crop, responsible for significant damage to seedlings and even mature trees (Ogedegbe and Ogwu 2015). These termites attack young plants at the collar, taproot, and stem base, mainly in the dry season, when the infestation can go unnoticed until plants suddenly start to wilt (Wood and Lass 1985, 424). If the plants are not killed, growth and yield can be reduced as a result of reduced translocation of water and nutrients from the roots to the leaves. Some termites establish colonies within the trunk or larger branches, feeding on the wood, hollowing out the trunk or branch, and filling it up with wet soil. The soil dries out and becomes hard, thereby keeping the damaged plant upright (Ambele et al. 2018).

As if direct damage by the termites wasn't bad enough, they can also carry spores of the black pod pathogen, with the prospect of even more damage and yield loss (Ogedegbe and Ogwu 2015). And to make matters worse, their direct damage to the cacao plant creates openings for fungi that just love to feast on wood (Vos et al. 2003).

Termite diversity and abundance, and the extent of damage they cause, are affected by a range of factors, including the practice of slash-and-burn agriculture and deforestation (Ambele et al. 2018). Forest clearing and the subsequent burning of felled trees can have adverse effects on populations of natural enemies of termites such as ants, leading to an explosion in termite numbers (Ambele et al. 2018). In Australia, clearing tropical savanna woodland led to a reduced abundance and diversity of termites, with a cascade of effects for water storage and retention in soils as the wet season ended (Dawes 2010). Fires can also affect termites. After a fire destroys much of their food supply, those termites that survive switch from feeding on dead trees to live trees, becoming pests in the process (Bandeira et al. 2003; Attignon et al. 2005).

Another factor affecting termite abundance and diversity is shade. In cacao, termite diversity was greater in shaded than unshaded systems, with shaded cacao harboring a greater diversity of nonpest species. Soil-feeding termites, so important in maintaining good soil conditions, are more vulnerable to reduced shade than wood-feeding termites, which are becoming increasingly important cacao pests. Research on cacao agroforests in Cameroon found that around 55% shade cover was best for balancing termite infestations and marketable yield (Felicitas et al. 2018).

How to Manage Termites in Cacao

One way of tackling termite trouble is to destroy the highly visible mounds. This approach is practiced in some Sub-Saharan countries, such as Ghana, Malawi, Uganda, and Zambia. Farmers dig out the entire mound until the queen and king are found and removed. Workers and soldiers are exposed to the sun as the mound is destroyed, and they desiccate and die rapidly. But all this hard work offers only a brief respite, because termite activity can return to normal pretty quickly. There is also the problem that not all termite nests are visible, and these tend to be missed (Nyeko and Olubayo 2005; Ambele et al. 2018).

Chemical control is used, but is expensive, labor-intensive, and can impact human welfare and the environment (Ambele et al. 2018). Using the toxic and deterrent properties of plants offers another, potentially less environmentally damaging means of dealing with termites. So, neem oil and extracts of neem leaves have been used to control termites in Nigeria and Ghana, while cacao farmers in Cameroon have used extracts of moringa (*Moringa stenopetala*) for the same purpose. But although there is a long list of plants and plant extracts used to control termites, this approach is beset with problems. For a start, efficacy is variable because most of the plant extracts are unstable, breaking down in sunlight. Then there is the issue of concentration of the active ingredient in the plant part or extract. This is usually a great deal lower than the concentration required to kill the termites, which means that huge quantities are required (Isman 2006; Ambele et al. 2018).

Using fungi to attack and kill termites has been shown to work in the laboratory, but their effectiveness in the field has been limited because termites can detect and avoid infected compatriots (Yanagawa et al. 2008; Ambele et al. 2018). Another possibility is to use the so-called attract and kill approach. First, you need to find some means of attracting the termite. Soil-dwelling insects such as termites use chemical and physical cues to help them find plant roots and to determine whether the plant is a suitable host (Barsics et al. 2016). Carbon dioxide released from plant roots allows insects, including several termite species, to find roots in the soil (Johnson and Gregory 2006). In the "attract and kill" approach, carbon dioxide is encapsulated with a toxic agent along with something to stimulate the insects to feed. This strategy has been used successfully in Europe to control a range of soil pests (Schumann et al. 2013) but has yet to be tested against termites.

Controlling any of the pests and pathogens we've looked at in this book is fraught with difficulty, and termite control is no exception. Ultimately, cacao farmers need to be made aware of the pest problems, which means that dissemination of up-to-date information on management practices is essential.

As we've seen time and again in this book, cacao is a shade-loving tree. In fact, shade has a profound effect not just on the tree, but also on the way it responds to the environment and other organisms. So, before we look at climate change and how it affects cacao, let us explore, briefly, the interaction of cacao, shade, and agroforestry.

13

Cacao, Shade, and Agroforestry

As we've seen, in its natural habitat, cacao is an understory tree, growing in the shade of considerably larger trees. It's no surprise, therefore, to learn that 70% of the world's cacao is grown under some level of shade (Gockowski and Sonwa 2011). The shade comes mostly from native forest trees, thinned out to provide space for cacao seedlings to be planted, or, to a lesser extent, from trees specially planted to provide shade. Banana and/or plantain is commonly used to provide shade to young cacao trees; as the trees mature, the native trees that are retained and new, exotic planted trees provide not just shade, but can also be used to provide extra income (Asare and Sonii 2010). This mixture of shade trees and shrubs creates a three-tier canopy, one level below the cacao trees and two levels above them. The result is a multispecies system similar in structure and function to a forest—an agroforest (Ruf 2011). In fact, cacao researchers have reported six general types of cacao agroforestry systems, including full sun cacao, cacao under diversified shade, and cacao under specialized shade (Johns 1999; Moguel and Toledo 1999; Rice and Greenberg 2000).

One of the criteria for differentiating among cacao agroforestry systems is the amount of light transmitted through the upper tiers of the canopy to the cacao trees. This, in turn, depends on the density and spatial structure of the shade trees in the system. Of the light received by a plant, only part is used in photosynthesis. This light is known as photosynthetically active radiation and is measured as $\mu mol/m^2/s$. According to some researchers, to grow and develop normally, a cacao tree requires 400 μmol/m²/s (Avila-Lovera et al. 2016). A study of different agroforestry systems in the Colombian Amazon found that the system with least shade provided light levels of 1400 μmol/m²/s, while the most shaded system had light levels of 680 μmol/m²/s. This is considerably more than the 400 μmol/m²/s required by cacao trees for their normal growth and development. The researchers think this finding is important, because ensuring that cacao trees have optimum light is one of the main reasons for reducing numbers of shade trees in cacao agroforests (Salazar et al. 2018).

Cacao is a shade-tolerant species and as such, it is accustomed to living with low light levels. As a result, its rates of photosynthesis reach saturation at relatively low levels of light. In fact, some researchers have shown that rates of photosynthesis in cacao leaves start to decline when light levels go above 200 μmol/m²/s, with the decline accelerating at light levels above 500 μmol/m²/s (Raja Harun and Hardwick 1988). Although cacao leaves can adapt to higher light levels, they seem to pay the price later on, because the photosynthetic machinery gets damaged. This means that rates of photosynthesis decline much more rapidly in these leaves as they age, compared with cacao leaves growing under shade (Miyaji et al. 1997a, 1997b).

So, how does shade in cacao agroforests affect cacao yield? Well, in Cameroon, high yields were associated with between 40% and 50% shade in 15-year-old trees in a semicomplex cacao agroforest, while in Indonesia, highest yields were obtained in a simple agroforestry system providing 30% to 40% shade (Bisseleua et al. 2009; Gras et al. 2016). Some workers report that low shade, providing between 0% and 25% cover, saturates cacao trees with light and leads to stress. Here, yields are not necessarily increased, especially in the longer term, say 20–25 years (Abou Rajab et al. 2016). This suggests that cacao agroforestry systems can provide better conditions for producing cacao than either cacao monocultures or forests with dense canopies (Tscharntke et al. 2011).

Growing cacao in the shade of agroforestry systems seems to have a lot going for it. Research suggests that not removing shade once cacao trees have matured contributes to improved conservation of biodiversity, by providing buffer and refuge zones for wildlife (Perfecto and Vandermeer 2008; Saj et al. 2017). The biodiversity benefits are due, in part, to the high diversity of tree species in cacao agroforests. This diversity ranges from 280 tree species reported for some Brazilian cacao agroforests, to 237 species in Côte d'Ivoire and 53 species in Costa Rican systems (Deheuvels et al. 2012; Sambuichi et al. 2012; Tondoh et al. 2015). Recently published work from the Colombian Amazon found 127 tree species in cacao agroforests (Salazar et al. 2018). But this high level of biodiversity was being eroded, with trees being removed to reduce shade for cacao cultivation. Nevertheless, the authors reckon that, providing they are developed and managed appropriately, Colombian cacao agroforests can still meet biodiversity conservation targets.

Tropical forests are one of the main sinks for carbon, and their loss through slash-and-burn practices to open up land for cultivation, logging, and other human activities leads to losses of carbon through vegetation loss, leaching, and soil erosion (Hosonuma et al. 2012). Although large areas of forests are lost to make way for perennial plantation crops, including cacao, the development of cacao agroforestry systems can help to maintain other ecosystem services, including soil fertility, erosion management, and carbon storage, while remaining productive in the long term (Jagoret et al. 2012; Mortimer et al. 2018).

But with demand for cacao beans growing at around 1% annually, there is a push to intensify cacao cultivation in order to maintain a secure supply (ICCO 2014). In the past, intensifying cacao production has been achieved by removing shade, resulting in reduced biodiversity (Ruf 2011). Today, the move to further intensify cacao production is likely to come not just from removing shade, but also from the use of improved genetic material and agrochemicals, including inorganic fertilizers. There is a worry that this will exert a negative impact on rural cacao communities and the conservation of natural resources (Vaast and Somarriba 2014). Minimizing any negative impact will require reconciling the two extremes—use of high external inputs to increase cacao yield versus maintaining appropriate shade levels and preserving species richness (Steffan-Dewenter et al. 2007).

Benefits of Shade and the Perils of Removing It

The push to intensify cacao production by removing shade seems to have its roots in a couple of studies in West Africa and one in Brazil, which reported increased yields following shade removal. In the first study, researchers in Ghana found that cacao yields were reduced by 22% under moderate shade, and by 50% under heavy shade (Ahenkorah et al. 1987). This study, however, used just one species of shade tree, the fast-growing pioneer tree *Terminalia ivorensis*. The second study, carried out in Côte d'Ivoire over just three and a half years, found a 2.5-fold increase in production of healthy pods under full sun compared with shade (Lachenaud and Mossu 1985). In the Brazilian work, total removal of shade, coupled with fertilizer use, led to a doubling of yield (Johns 1999). These three studies were conducted between 20 and 30 years ago, and some workers reckon that the results may not be relevant to the newer cacao germplasm in use today (Vaast and Somarriba 2014). Unfortunately, in the intervening period, the vast majority of cacao genetic trials have used full sun, with no shade treatment. Perhaps the time has come to redress the balance.

But although pod yields of mature cacao trees can be increased by removal of shade trees, the increases are only short-term. This was shown clearly in the results of a 14-year experiment carried out in Ghana and published in 1974 (Ahenkorah et al. 1974). In this work, treatments were applied to 10-year-old trees, and large yield increases were obtained when shade was removed. The largest yield increases were obtained when fertilizer was applied, but even so, yields began to decline after about 10 years, while yields of trees under shade and given fertilizer continued to increase (Ahenkorah et al. 1974).

Generally, in unshaded situations, when the trees get beyond 25 or so years old, pod yields fall, and insect pests become problematic. These pressures often lead to farms being abandoned, and new areas of forest used to start again. This has been called a short-term boom-and-bust cycle and has been common practice in cacao cultivation (Clough et al. 2009). Unfortunately, the pest and pathogen problems associated with the short-term boom-and-bust cycles can lead to a long-term boom-and-bust cycle with countrywide consequences. This happened in Brazil in the late 1980s, when pathogen problems turned the cacao boom into a cacao bust, resulting in a huge slump in production. A similar fate befell Malaysia a few years later, courtesy of cocoa pod borer (Bos et al. 2007).

Cacao grown in shaded agroforestry is said to be less affected by insect pests, although the relationship between cacao agroforests and pests is complex (Mortimer et al. 2018). According to the *insurance* hypothesis, organisms are less likely to attain pest levels in complex assemblages of predators, because the high levels of species richness help with increasing resilience to disease and pest outbreaks (Rice and Greenberg 2000). So, shade trees in cacao agroforests can encourage ant populations capable of protecting cacao trees from attack by *Helopeltis theobromae,* and they are also associated with creating an environment not to the liking of many pests, which tend to prefer lots of sun (Schroth et al. 2000; Babin et al. 2010). These sun-loving pests include thrips, mirids, and the leaf miner *Leucoptera meyricki.* In fact, getting rid of ants can be bad news. Researchers in Indonesia found that excluding ants from cacao agroforests with different shade levels reduced cacao yields because of increased pest levels in all but the most highly shaded system (Gras et al. 2016). Birds and bats also eat insect pests, and excluding them from cacao agroforests can lead to increased pest numbers and greater damage to cacao.

The increased humidity associated with shaded cacao could favor pathogens, although this seems to depend on the type of shade. Cacao growing under the shade of a canopy of diverse native trees actually suffered significantly less from black pod than cacao under planted shade trees. The researchers who carried out this work suggested that this might have been due to the greater abundance and diversity of microbial antagonists present under natural shade (Arnold and Herre 2003). Shade also enhances the diversity of endophytes, those microbes that live quietly and unseen within cacao leaves, some of which just love to antagonise the black pod pathogen (Schroth et al. 2000). Interestingly, in the study of cacao agroforests in the Colombian Amazon, researchers found that optimizing the spatial organization of shade trees was more effective in controlling frosty pod rot than reducing the density of shade trees (Salazar et al. 2018).

But little in life is clear-cut and so it is with shade and cacao diseases, because high humidity in shaded cacao agroforests can increase disease problems. In southern Cameroon, for example, humidity is high and so are the ravages of the black pod pathogen *Phytophthora megakarya*. Farmers are encouraged to reduce humidity by pruning and optimizing the spacing of shade trees, although, in reality, fungicide applications are important in tackling the pathogen (Niether et al. 2018).

A recent study in Ghana looked at cacao production and shade tree management along a climate gradient within the cacao belt in Ghana (Abdulai et al. 2018). The researchers identified two agroforestry shade systems: a medium-shade system and a low-shade system. The medium-shade system was dominant in the dry areas, which are less suitable climatically for cacao, while the low-shade system was favored in wetter regions, which were a better climatic fit for cacao. The two shade systems had different effects on cacao yields. So, medium shade exerted a positive effect on yields in the drier regions, but decreased yields in the wet regions. By contrast, yields were significantly higher in the low-shade systems, but only in the wet regions. Importantly, farmers in wet areas considered black pod the main reason shade needed to be reduced. Not surprisingly, the researchers who carried out this work suggested that the specific climatic conditions of a region should be taken into account before decisions are made regarding shade management (Abdulai et al. 2018).

Returning to pests, briefly, the issue of why cacao should suffer so badly from long-term pest problems is not known, although part of the reason may relate to how pests respond to the amount of crop planted. It has been suggested that as the area of cacao planted increases, so too will the diversity of pests feeding on it. Pests are also likely to be attracted to a high density of cacao, which would provide them an abundant food supply concentrated in one area and make it easier for them to move between plants. As if all of this weren't enough, the cacao tree becomes more susceptible to pathogens when it is stressed, which is what happens when there isn't enough shade. To compound matters, weeds thrive as shade is reduced, and in addition to competing with the cacao tree for water and nutrients, they also provide a safe hideaway for pathogens and pests (Clough et al. 2009).

But the benefits of shade trees for cacao run to more than lessening the impact of diseases and pests. Leguminous shade trees fix nitrogen, converting atmospheric nitrogen into organic forms that the plant can use, thereby reducing the need for fertilizer application. This can be important, since nitrogen fertilizer applied to soil can be converted by soil bacteria to the powerful greenhouse gas nitrous oxide. In Indonesia, nitrogen fertilizer applied to cacao agroforests leads to very high nitrous oxide emissions (Veldkamp et al. 2008). In case you're wondering, nitrous oxide is also emitted from shaded cacao agroforests, but at lower rates.

Shade trees can also help to protect cacao against drought. Under shade, temperatures are lower and humidity higher, which reduces water stress in the cacao trees. Research in Sulawesi has shown that the increased canopy cover from shade trees allowed cocoa trees to take up more water from the soil; these trees also grew larger stems and leaves (Bos et al. 2007). But there is another side to this story. The canopy of shade trees will intercept rainfall, reducing the amount of water reaching the soil. The cacao will then need to compete with the shade trees for soil water, which could lead to reduced growth and yield. It doesn't need to be like this, however, because positive effects can be achieved with certain species of shade trees. Here, there would be complementary resource use, rather than competition for resources that the shade tree would win. Ultimately, how well the cacao tree does under shaded systems will depend on the types of shade trees providing the cover, and environmental factors such as drought severity (Bos et al. 2007; Tscharntke et al. 2011).

Pruning Shade Trees?

As we've seen, agroforestry systems, with their canopy of shade trees, provide a buffer against extreme climatic changes, and in so doing, reduce stress for the cacao tree. But, as mentioned above, the dense canopies in such systems will reduce rainfall reaching the soil and will also reduce light available for photosynthesis, both of which would limit cocoa growth and yield. What if the canopy provided by the shade trees could be reduced just enough to continue to provide buffering against climatic extremes, while allowing adequate rain and light to get through to the understory cacao? In other words, might pruning of the shade trees offer a means of providing cacao with favorable growth conditions? This question was examined by researchers working in Bolivia, who found that by pruning shade trees, they could alter the microclimate in favor of cacao, while maintaining the diversity of shade trees (Niether et al. 2018). The problem is that although farmers are used to pruning their cacao trees, they rarely prune shade trees, partly because this practice is unfamiliar to them and also because they don't have the necessary tools for the job. If they are planning to prune shade trees, the timing and intensity of the pruning depends on where they are, because major considerations are seasonal changes in temperature and rainfall. So, in some regions, farmers might need to increase rainfall getting through to the soil during the dry season, without reducing the buffering function of the canopy. In other areas, such as Alto Beni in Bolivia, pruning of cacao and shade trees would be carried out at the end of the dry season, because temperatures thereafter are higher than at other times of the year (Niether et al. 2018).

Lessons from the Past

We've already seen that part of global cacao production comes from complex cacao agroforestry systems. In some parts of the world, these systems can be many decades old. In Cameroon, for example, 40% of cacao agroforests are more than 60 years old, and they are still being used by farmers (Jagoret et al. 2011, 2018). In an interesting study published in 2018, Patrick Jagoret and colleagues examined how 30 cacao agroforests evolved over the decades. They identified five different trajectories along which the agroforests evolved, leading to present-day agroforestry structures comprising, for example, low or high cacao tree densities, and low to high yields. The results show that these systems are resilient and flexible, features related to their species complexity. In these systems, cacao can be produced much longer than is possible under cacao monoculture, where cultivation is usually abandoned and replaced either by other crops or by new areas of forest sought for cacao production (Jagoret et al. 2018).

Ultimately, cacao farmers, like everyone else, need to make a living. They need to know that growing cacao in an agroforestry system is going to pay. Research published in 2017 suggests that cacao agroforestry in Ghana is profitable. A study of 200 cacao farmers from the Western region of Ghana found that medium-shade agroforestry was more profitable than no-shade, low-shade, and heavy-shade systems. The researchers suggest that promoting medium-shade cacao agroforestry would be the right policy to ensure the welfare of cacao farmers and enhance environmental sustainability (Nunoo and Owusu 2017).

It is now well established that shade trees in cacao agroforests are important for farmers' livelihoods and the conservation of natural resources. It stands to reason, therefore, that there should be a detailed assessment of the long-term effects of shade removal on cacao yield, taking into account differing socioeconomic and ecological conditions (Vaast and Somarriba 2014). This is especially important as the world's climate changes, because, as we will see in the following chapter, cacao grown in agroforestry systems might be better able to cope with the changing environment.

14

Cacao in a Changing Climate

Earth, our wonderful blue planet, has endured many changes in temperature during its 4.5-billion-year existence. Steamy tropical climates have given way to ice ages as part of the planet's long-term changes in weather patterns and average temperatures (Met Office, n.d.). There have been many ice ages during the last 2.6 million years, with warmer, interglacial periods between them. We are in the middle of an interglacial period now, the last ice age having ended around 11,500 years ago. Since then, the average global temperature has been about 14 °C. Recently, however, things have begun to change, with a significant increase in global temperatures over the last century (Henson 2008). So, what exactly is going on?

Most of the changes in Earth's climate in the past have been the result of very small variations in our orbit, which altered the amount of energy we received from the sun (NASA Global Climate Change, n.d.). The recent increases in global temperature are different because it is considered by the vast majority of scientists to be the result of human activity since the middle of the twentieth century (IPCC 2014). We know this partly as a result of data collected over the past 300 years or so (Henson 2008, 3), and more recently because huge technological advances have enabled scientists to collect information about the earth and its climate on a global scale. Jumping out of this huge volume of data are the signals of a climate that is changing rapidly (NASA Global Climate Change, n.d.).

Climate Change: It's a Gas! Gas! Gas!

Climate change is never out of the news, and rightly so, because it is a huge threat to our planet. As a result, most of us will be familiar with the term *greenhouse gases*. But what does this term mean?

To answer this, we need to go back to the 1820s and the French mathematician and physicist Joseph Fourier. He was interested in Earth's energy balance, and according to his calculations, there was a massive temperature difference between the Earth we know, with its atmosphere, and Earth without an atmosphere. He knew that the energy arriving at Earth's surface as sunlight must be balanced by energy returning to space. If Earth were an airless planet, the sun's radiation would heat the surface up like a furnace during the day, whereas at night, temperatures would drop like a stone. On Mars, which has a thin atmosphere comprising mostly carbon dioxide, nitrogen, and argon, temperatures can be as high as 20 °C or as low as –153 °C (NASA Science, Solar System Exploration, n.d.). But of course, Earth's temperature does not yo-yo between such extremes; instead, it averages out to a more accommodating 14.4 °C. Fourier compared Earth's atmosphere to a greenhouse, trapping heat and preventing it from escaping into space. This comparison gave rise to the idea of the "greenhouse effect" (Henson 2008, 20–21).

But Earth's atmosphere does not trap heat like a greenhouse. Of the sun's rays reaching Earth, mostly in the form of shortwave radiation, roughly 30% is reflected back to space by clouds, dust particles, and the ground, about 20% is absorbed by clouds and water vapor in the atmosphere, and nearly half is absorbed by Earth's surface—oceans, land, forests, and so on. Radiation leaves Earth as infrared or longwave radiation, with some escaping directly to space through the atmosphere, but most is absorbed by clouds and greenhouse gases. Some of this radiation is reflected back to Earth's surface and some escapes to space. The result is a favorable balance between incoming radiation from the Sun and radiation escaping to space from the combination of Earth's warm surface and a cooler atmosphere (Henson 2008, 22–23).

Nitrogen and oxygen make up 98% of the air we breathe, but they don't absorb radiation from Earth very well. Of the remaining gases in air, carbon dioxide is much better at absorbing radiation and does so out of proportion to the amount of it present in the atmosphere. Carbon dioxide, along with water vapor and methane, are greenhouse gases. Things get a lot cooler as you move up through the atmosphere from the ground, and at higher altitudes, the greenhouses gases like carbon dioxide are also cooler (Henson 2008, 24–27). Because of this, they radiate less energy to space than Earth's surface would, thereby keeping more heat in the atmosphere and making the planet habitable (Henson 2008, 24–27).

But what would happen if the amounts of any of these greenhouse gases in the atmosphere were to increase? Well, they simply get in each other's way, blocking radiation to space, and hey presto, the atmosphere heats up. This increase in temperature leads to greater evaporation of water from oceans and lakes, and as water vapor in the atmosphere increases, atmospheric temperature increases further. Increasing temperature also causes sea ice to melt, and because ice reflects radiation back to space, less ice means less sunlight reflected to space (Henson 2008, 24–27). Result? Yes, the atmosphere heats up.

Greenhouse Gases: The Main Offenders

Okay, so increasing the amount of greenhouse gases in the atmosphere heats things up. But are all greenhouse gases the same, or are some worse offenders than others? Let's start with carbon dioxide. Animals, including we humans, take in oxygen when they breathe in and expel carbon dioxide to the atmosphere when they breathe out. This is the physical process of breathing, but is often called respiration. In fact, respiration is a cellular process used by all organisms. In respiration, organisms use food such as carbohydrates and fats, together with oxygen, to produce the energy required to maintain them and to grow. Incidentally, not all organisms use oxygen in respiration, which is known as aerobic respiration—some use other substrates, such as sulfur, in which case the process is called anaerobic respiration. The bottom line is that in aerobic respiration, oxygen is used, and carbon dioxide is released as a by-product. Carbon dioxide is also produced when organisms decompose and, of course, when we burn fossil fuels such as coal and gas. Plants take up carbon dioxide from the atmosphere and use it in photosynthesis, converting it to sugars. In fact, plants and the oceans absorb large amounts of carbon dioxide, without which levels would rise even more steeply than they have been. Carbon dioxide levels in the atmosphere remained at around 280 ppm for centuries, but things started to change with the advent of the Industrial Revolution. Since then, carbon dioxide levels have been increasing at a rate unprecedented for more than 800,000 years (Met Office, n.d.; Farand 2018). When I was an undergraduate in the 1970s (1975–1978), the carbon dioxide concentration in the atmosphere was just over 330 parts per million (ppm).

In April 2018, just 40 years after I finished my degree studies, the National Oceanic and Atmospheric Administration's observatory at Mauna Loa on Hawai'i found that carbon dioxide concentration in the atmosphere reached 410 ppm across an entire month for the first time since it had started its measurements. As of October 2019 it stands at 412 ppm (Farand 2018).

Methane is produced naturally, in peat bogs, for example, and by human activities, the latter including rice cultivation, decomposition of waste in landfill sites, and belching cows. Although it is much less abundant in the atmosphere than carbon dioxide (<2 ppm compared with >400 ppm for carbon dioxide), it is a far more active greenhouse gas (Beerling 2007, 155). As Earth's climate started to warm up after the end of the last ice age, methane concentration started to increase. So, a warmer climate led to more methane in the atmosphere. But remember that methane is a greenhouse gas, so the increased methane, in turn, led to further increases in atmospheric warming (Beerling 2007, 155–157).

Why would a warming climate increase atmospheric methane? To answer that, let's visit the swamp, for it is here that anaerobic microbes obtain the energy they need by feasting on organic matter. A by-product of this guzzling of organic material is the release of methane, which bubbles to the swamp surface. Just how much methane bubbles up from the swampy depths depends on climate because temperature, for example, will influence the activity of the microbes and the rate at which they decompose organic matter in the swamp (Worden et al. 2017). Methane concentration in the atmosphere has increased sharply since 2006, with increased emissions from both biogenic sources such as swamps, ruminants, and rice paddies, and the burning of fossil fuels. NASA scientists have calculated that the increased methane comes mostly from biogenic sources, which contributed around 84% of the increase, with the remainder coming from fossil fuels (NASA Global Climate Change, n.d.; Royal Society/National Academy of Sciences 2014).

We don't want methane levels in the atmosphere to increase further, and so it is unsettling to discover that sediments on the ocean floor at the margins of continents, as well as permafrost areas at northern latitudes, are a massive store for the gas in the form of methane hydrate. This is a frozen, naturally occurring, highly concentrated form of methane, which degrades as temperatures increase, releasing the gas. The worry is that release of methane from just a small part of this huge store could exacerbate climate change (Ruppel and Kessler 2017).

As greenhouse gases go, water vapor is not particularly potent, but it makes up for this deficit by its sheer abundance in the atmosphere. As temperatures rise, more water vapor is released from oceans and lakes, adding to the cycle of global warming. Carbon dioxide, methane, and water vapor are major players in global climate change, but there are minor players too. These include nitrous oxide, which is produced by soil cultivation practices such as the use of fertilizers, the combustion of fossil fuels, and the burning of biomass, including wood (Henson 2008, 22–23; NASA Global Climate Change, n.d.).

How Do We Know Earth's Climate Is Changing?

Our climate is warming up. In fact, Earth's average surface air temperature has increased by about 0.8 °C in the past 118 years, although a large proportion of this increase has occurred since the 1970s (Royal Society/National Academy of Sciences 2014). Estimates of temperature change from a range of locations and sources reveal that 1983–2012 was probably the warmest 30-year period in the past 800 years. But these changes are set to continue, especially if we do nothing to reduce emissions of greenhouse gases. If we take no action, the projection is for further global warming of between 2.6 and 4.8 °C (Royal Society/National Academy of Sciences 2014). It's hard to imagine that temperature increases of just a few degrees, even 4.8 °C, can be important. But consider this—global average temperature during the last ice age was only about 4–5 °C colder than today. During that period, huge swathes of the planet were covered by massive ice sheets. So, the increases projected will be associated with changes in temperature and precipitation, both locally and regionally, as well as with an increased frequency of extreme weather events (Royal Society/National Academy of Sciences 2014).

Although extreme weather events occur as part of natural variability in Earth's climate, they usually occur rarely. The oceans have a huge role to play in our weather and climate, mostly via a set of ocean-atmosphere cycles. These cycles are linked to high and low pressure centers over various parts of the world, and they alternate between two modes, giving rise to repetitive weather patterns, including drought and excessive rainfall (Henson 2008, 118–119). The most important of these cycles is the El Niño–Southern Oscillation (ENSO), the cycle of warm and cold sea surface temperatures that occurs in the tropical central and eastern Pacific Ocean. The warm phase of this cycle is called El Niño and is associated with a band of warm ocean water that develops in the central and east-central Pacific, including off the

Pacific coast of South America. Lasting typically one or two years, it is accompanied by high air pressure in the Western Pacific and low air pressure in the eastern Pacific. The cool phase of the cycle is known as La Niña and is associated with high air pressures in the eastern Pacific and low pressure in the west. When neither of these two phases is in progress, the Pacific Ocean is neutral.

When El Niño is operational, the chances of droughts occurring across Indonesia, Australia, India, Southeast Africa, and northern South America are increased, while wetter than normal conditions occur in Ecuador, Peru, and southern Brazil during the period from December through to February (Henson 2008, 118–119; ICCO 2010). El Niño brings cool, moist winters to south and southwestern USA, and mild, dry winters to the northern parts of the USA and in Canada. It also increases the odds of hurricanes developing in the Atlantic and reduces the chances of them occurring in parts of the north Pacific. By and large, La Niña does the opposite (Henson 2008, 118–119).

Between 1950 and 2010, there were 18 El Niño and 13 La Niña events. Scientists regard the El Niño events of 1982–1983 and 1997–1998 to be the most severe recorded during this period. It is easy to see why they came to this conclusion when you realize that during the 1982–1983 event, Ecuador and Peru received about seven years' worth of rain in just four months, while Indonesia and Malaysia suffered from severe droughts and uncontrollable forest fires (ICCO 2010).

As the climate warms, these short-term and regional variations are expected to become more extreme, with El Niño events predicted to be amplified by climate change. In fact, heavy rainfall and snowfall events, which increase the risk of flooding, as well as heat waves, are becoming more frequent (IPCC 2014; NASA Global Climate Change, n.d.). In the UK, the Met Office has predicted that northern parts of the country will see more frequent high-intensity rainfall events, where 10 mm or more of rain falls in an hour, with less rain falling during the summer (Met Office, n.d.). In the US, more winter and spring precipitation is projected for northern areas, and less for southern areas, over the rest of this century (NASA Global Climate Change, n.d.). Climate change is expected to hit Africa hard, with an increase in severe droughts, floods, and storms that are likely to threaten human health and the economies of many countries on the continent. Africa is already the hottest continent but is expected to get warmer at a faster rate than the global average (Royal Society/National Academy of Sciences 2014).

How Is Climate Change Likely to Affect the World's Cacao-Growing Regions?

Nearly three-quarters of world cacao production is concentrated in West Africa, making the industry highly vulnerable to any deterioration in climatic suitability for growing the crop. Since it is an understory tree, it is sensitive to prolonged drought and high temperatures, and unfortunately, cacao in the region is already subject to significant drought risk (Schroth et al. 2017). The climate of West Africa, particularly the Sahelian region, is one of the most variable on the planet, and this variability has increased during the twentieth century. The 1930s into the 1950s saw unusually heavy rainfall in the region, while the subsequent 30 years brought periods of extended drought (Brown and Crawford 2008). Drought years can affect cacao yields, and this was particularly noticeable during the severe El Niño years of the 1980s (Brown and Crawford 2008). It has been estimated that El Niño events reduce world cacao production by an average of 2.4%. By contrast, La Niña events have no significant impact on world cacao production (ICCO 2010).

It is predicted that temperatures in West Africa will increase progressively over the next 50 years by as much as 2 °C. This increase in temperature is expected to increase evapotranspiration, the loss of water from the soil and plants, leading, in turn, to increased water demand by cacao trees. If cacao trees need to take up more water, but less is available because it has evaporated to the atmosphere, trees could easily become water-stressed. This would be a real problem during the dry season, especially in years when El Niño is active, which would dry things out even more (Läderach et al. 2013; Ruf et al. 2015). It seems, therefore, that the availability of water during the dry season might be the key factor in determining whether the West African climate will continue to be suitable for growing cacao in the future.

According to the most recent predictions, however, water availability is not likely to be as important as previously thought. These predictions suggest that increased evapotranspiration, and the resulting increased demand by cacao trees for water, is likely offset by greater rainfall and a shorter dry season (Läderach et al. 2013). Although these predictions affect much of the West African cacao belt, there are exceptions—those areas at the northern edge of the cacao belt, encompassing eastern Côte d'Ivoire, Nigeria, and Cameroon— where it is likely to get drier. What the authors of this study are saying is not that cacao in West Africa won't suffer from drought in the future; just that it might not be much different from the current situation.

But before you start thinking that nothing much is going to change, we have not dealt with temperature yet. As we saw in chapter 3, the cacao tree likes it warm but not too hot. Its limit is 38 °C, beyond which its physiology suffers, with rates of sugar-producing photosynthesis falling (FAO 2007; Schroth et al. 2016). Currently, maximum temperatures for most of the West African cacao belt are less than 35 °C, and 36 °C tends to be reached only at the northern edge of the cacao belt, where the forest meets the savanna. With climate change, it is predicted that by 2050 maximum temperatures of 34–36 °C will be more common within the cacao belt. Temperatures in excess of 36 °C, now confined to savanna areas in West Africa, will affect northern areas of the cacao belt, especially in Togo, Côte d'Ivoire, and Nigeria. Although the physiological limit of cacao, 38 °C, is not expected to be reached for the cacao belt by 2050, it will get close in some areas close to the main cacao belt, especially in Guinea, Sierra Leone, and Liberia, which are influenced by the hot savanna of Guinea; 38 °C might also be experienced in Togo and Nigeria, which of course are in the main cacao belt (Schroth et al. 2016).

What exactly do these projected increases in temperature mean for cacao production in West Africa? Well, by the 2050s, cacao in most of the cacao belt would be exposed to temperatures currently only experienced at the northern edges, where the forest transitions into savanna. If the optimum temperature for cacao is reached and photosynthesis starts to fall, this will, in turn, reduce growth of the tree and its yield. But cacao is not the only tree likely to be affected by climate change. Companion trees, used to provide shade for cacao trees, might also be affected. In the future, such trees will need to be selected carefully for their ability to tolerate elevated temperatures, because shading is likely to be important for cacao growing in a warmer climate. Shading can reduce the temperature of cacao leaves by up to 4 °C, which would help greatly if temperatures are nearing cacao's physiological optimum (Almeida and Valle 2007; FAO 2007; Schroth et al. 2016).

According to a study published in 2017, however, shading might not be the answer. The authors looked at what was likely better for cacao under climate change: growing the crop as a monoculture without shade, or growing it as part of shaded agroforestry (Abdulai et al. 2017). The study concluded that monocultures were better for cacao under climate change, because in shaded agroforestry, there is competition for water between cacao and the shade trees. These conclusions have been called into question by other researchers, who say that the experimental setup was not appropriate for the conclusions drawn. As they point out, cacao trees don't like full sun and suffer physiological stress when grown without shade. Such stress can lead to increased attacks by pests; in the past, this has resulted in abandonment of cacao cultivation (Wanger et al. 2018). Shade trees are also important in protecting against soil erosion, they provide a litter layer, and they could be useful in providing more effective management of pollinating insects (Wanger et al. 2018). But, in much of West Africa, the traditional practice of growing cacao under the shade of remnant forest trees is being replaced by cacao cultivation under low or zero shade. According to some researchers, this makes cacao production highly vulnerable to the risks posed by climate change (Schroth et al. 2017).

Adapting to Climate Change

Adapting to climate change is often viewed as a process intended to reduce the losses farmers and growers are likely to face, and in so doing, to reduce any hardships. Typically, these actions tend to be focused at a relatively local level, contrasting with climate change mitigation, which tends to be seen as a global activity. So far, cacao has not been the subject of major efforts to reduce its vulnerability to climate change, but that appears to be changing. It has been suggested that, rather than adopting a local approach, adapting cacao for climate change should be carried out at a regional level, emphasizing the projections that different areas of the West African cacao belt are not all equally vulnerable to climate change. A study published in 2017 (Schroth et al. 2017) divided the West African cacao belt into zones of vulnerability to climate change, identifying actions appropriate for each zone, providing a sustainable development approach to adapting to climate change. In this study, Zone 1 encompasses an arc of land stretching along the Guinea coast from Sierra Leone through southwestern Liberia and southern Côte d'Ivoire to eastern Ghana, with a smaller arc covering southern Cameroon. This zone includes some of the world's most important cacao-producing areas, and here, the climatic suitability for cacao farming is not expected to change much by the 2050s. In Zone 2, which comprises the northern parts of the cacao-growing areas of Liberia, Côte d'Ivoire, Ghana, and Cameroon, as well as most of the cacao-growing areas of Nigeria, Togo, and Guinea, climatic suitability for cacao is projected to decline slightly (Schroth et al. 2017). Worst affected are the areas covered by Zone 3, including northeastern Côte d'Ivoire, part of adjacent Ghana, northern and western fringes of the Nigerian cacao belt, and the northeastern reaches of the cacao belt in Sierra Leone. Here, climatic suitability for growing cacao is expected to fall to such an extent that it seems unlikely that these areas could continue to grow cacao. The authors of the study suggest that dividing the cacao belt of West Africa in this way highlights areas such as Zone 3, where, in the face of deteriorating climatic suitability for cacao, farming systems would need to be adapted, including growing crops better suited to the newer climatic conditions. For this plan to be implemented successfully, however, actions will need to be coordinated at a national level, preferably integrated across the region (Schroth et al. 2017). Cacao farmers will need to adapt, but getting them to embrace the required changes may require governments to provide positive incentives for change.

Three-quarters of Brazil's cacao is produced in Bahia, in the northeast of the country. Here, it is grown in agricultural systems known as cabrucas, under the shade of large trees, a mixture of species native to the Atlantic rainforest and species introduced for food and timber. It is thought that this kind of diversification increases farmers' resilience to extreme climate events (Gateau-Rey et al. 2018). This might be put to the test, since some studies predict an increased frequency of strong ENSO events, which, as we've seen, are responsible for droughts and flooding in tropical regions. Such events caused droughts in northeast Brazil in the 1980s and 1990s, with effects on agricultural yields and forest cover. More recently, from October 2014 to May 2016, a strong ENSO event caused severe droughts in northeastern Brazil. Despite this track record and the importance of cacao in Bahia, no field study had been undertaken to examine the effects of ENSO-related drought on Brazilian cabrucas (Gateau-Rey et al. 2018).

This lack of field study was remedied by Lauranne Gateau-Rey and Edmund Tanner at Cambridge, who, together with colleagues in France, USA, and Brazil, examined the effects of the El Niño–related drought of 2015–2016 on agroforests in Bahia. They performed their measurements on 31 randomly chosen farms with traditional cacao agroforestry systems. It was the first time anyone had looked at the effects of severe natural drought on cacao trees before and after an ENSO event. What they found was worrying. The drought caused high mortality of cacao trees (15%), reduced yields by a whopping 89%, and increased infection by the dreaded witches' broom fungus. This is bad news because it suggests that Brazilian cacao agroforests are at risk from future climate events, which could reduce cacao yields. Such events might also affect other crops, which, like cacao, are grown slightly outside their normal climatic ranges, putting them at risk from more frequent, severe droughts (Gateau-Rey et al. 2018).

Cacao production in Indonesia is also likely affected by climate change. A report published in 2017 found that little cacao production in the region will remain unaffected by the changing climate. Low-lying areas in particular face an uncertain future because rising temperatures and highly variable precipitation levels will push climatic conditions beyond those currently experienced by cacao. Nevertheless, the report suggests that most cacao production in Indonesia will remain viable, although it will be necessary to increase the capacity to adapt to the changing climate (Bunn et al. 2017).

In the Caribbean, climate change is projected to make the climate warmer and drier. Lying fewer than seven miles off the northeast coast of Venezuela, Trinidad and Tobago has a tropical climate with distinct wet and dry seasons—the first five months of the year are dry, while the remainder of the year represents the wet season. According to a 2015 report produced by the International Center for Tropical Agriculture (CIAT), the effects of climate change on cacao in the Caribbean region are expected to be small. Although cacao production is not expected to be significantly affected by higher temperatures, the major risk factor is more likely to be increased severity of droughts in the dry season (Eitzinger et al. 2015; Farrell et al. 2018).

Deforestation, Climate Change, and Cacao

Forests are an important part of the carbon cycle of our planet. They take up carbon dioxide from the atmosphere in huge quantities, to satisfy the needs of photosynthesis, which then converts it into carbohydrates. A large proportion of these sugars are then used by individual trees to fuel their growth. Basically, forests act as sinks for carbon dioxide, removing it from the atmosphere and locking it up in biomass. When forests are cut down, there is a double whammy—not only can the felled trees no longer absorb carbon dioxide from the atmosphere; the carbon stored within them is released back into the atmosphere if they are burned or just left to rot.

It is estimated that between a quarter and a third of anthropogenic carbon emissions—carbon dioxide released to the atmosphere from burning fossil fuels and trees, for example—is taken up by tropical forests (Mitchard 2018). Although tropical forests are important sinks for atmospheric carbon dioxide, they are variable; that is, they don't always act as sinks and can, in fact, act as sources of carbon dioxide release to the atmosphere. This can happen in hot years when drought suppresses carbon dioxide–absorbing photosynthesis more than carbon dioxide–releasing respiration, resulting in the plants becoming net emitters of carbon dioxide (Mitchard 2018). Researchers are concerned that as climate change continues, with increasing temperatures and droughts, tropical forests will become a source rather than a sink for carbon, every year. This will not be helped by continued loss of tropical forests and will make it harder to limit global warming to below 2 °C (Mitchard 2018).

For the past 500 years or so, cacao cultivation, first in Latin America, then in Africa and Asia, involved moving to new sites once planted trees became too old. Forest would be cleared and new trees planted. In the early 1900s, the humid and subhumid belt of West Africa was a huge area of tropical rainforest. Today, much of this forest has been lost as a result of logging, slash-and-burn agriculture, and tree crop farming such as cacao cultivation. Between 1988 and 2008, global forest loss resulting from cacao planting was estimated at between 2 and 3 million hectares, equivalent to around 1% of total forest loss in that period (Resourcetrade.earth, n.d.). Cacao production is thought responsible for about 25% of historical deforestation in Côte d'Ivoire, and some 15% of forest loss in Ghana, and has also been responsible for forest loss elsewhere, including Cameroon, Peru, and Indonesia (Resourcetrade. earth, n.d.). In the next twenty years there is expected to be an increased demand for chocolate, driven largely by Asian demand (Ford et al. 2014). This increased demand is likely to lead to increased cacao production and could be damaging if the increased cacao plantings are in sun-grown systems rather than shade-grown systems under forest cover. Climate change could also lead to further loss of forest, if the reduction of climatic suitability for the crop means farmers expand into new areas, removing forest trees to plant cacao. Clearly, something needs to be done to safeguard forests. Protecting forests would also help cacao, given the many benefits associated with growing cacao under the shade of forest trees.

There is clearly a need to develop more sustainable cacao production practices, and the governments of cacao-producing countries, as well as those of consumer countries and the multinational chocolate companies are looking for solutions. So, the governments of Ghana and Côte d'Ivoire have national strategies for dealing with climate change that include the establishment of climate-smart cacao production systems, while the chocolate industry made a commitment in 2017 to put an end to deforestation and forest degradation in their supply chains, setting up the Cocoa and Forests Initiative (World Cocoa Foundation, n.d.). These are huge steps in the right direction; it is worrying, therefore, to hear reports in 2017 of illegal clearance of forests to make way for cacao plantations (Resourcetrade.earth, n.d.).

Recognizing the value of protecting forests in combating climate change, policies have been developed—known collectively as Reducing Emissions from Deforestation and Degradation (REDD+)—to provide a financial incentive to governments, businesses, and communities to maintain and increase, rather than reduce, forest cover (LSE 2018). Under these schemes, countries, communities, and landowners are offered incentives to protect forests in exchange for slowing down the rate of deforestation, and undertaking activities to promote reforestation and sustainable forest practices. The first REDD+ project in the Nawa region of Côte d'Ivoire was set up in 2017 by Mondelēz International to identify and monitor areas at risk of deforestation, and to create land-use plans (Cocoalife.org, n.d.). In neighboring Ghana, great progress has been made toward implementing REDD+ across the country, but although its REDD+ program has won praise, little information was available on how these climate change mitigation schemes were viewed by cacao farmers (Ameyaw et al. 2018). This lack of information was remedied by a study published in 2018, which examined the knowledge and perception of cacao farmers in Ghana on the potential impacts of climate change. The researchers found that even within the same geographical location, farmer knowledge and perceptions relating to climate change varied widely, presenting a considerable challenge to implementing REDD+. The study suggested that climate change mitigation strategies such as REDD+ may need to be more flexible and focus more on the human dimension, particularly where growing cacao is interwoven with forestry (Ameyaw et al. 2018).

Cacao Diseases and Pests under Climate Change

As we've seen, cacao is particularly vulnerable to climate change. Under future climate change scenarios, the crop will need to cope with higher temperatures, and in some areas, temperatures close to or exceeding its physiological limit. Higher temperatures would increase water loss by plants, and in addition, erratic precipitation events would lead to episodes of prolonged drought (Farrell et al. 2018). As well as effects on the cacao tree, there are also likely to be knock-on effects of the changed climate on pests, disease-causing organisms, and pollinators. Climate changes might, therefore, increase the prevalence of certain insect pests and microbial pathogens, and might hamper the effectiveness of pollinating insects.

Predicting the impact of future climate change on insects and pathogenic microbes is difficult because their biology and life cycles can be hugely affected by the environment. Nevertheless, you can begin to get an idea of how these organisms might be affected by looking at how they respond to particular changes in the environment, such as temperature and relative humidity. For example, insects are cold-blooded and so their rates of development are greatly affected by temperature. Increasing temperature will shorten the insect life cycle, but effects on insect populations are more complex. Higher temperatures might increase the number of generations produced in a year, improve their ability to overwinter, or increase their geographical range. The problem is that for some major cacao pests, such as pod borer, little is known (or at least published) about the effects of the environment on its life cycle. Even so, research has shown that second-instar nymphs love very high humidity, but dry out when humidity falls. In fact, temperature and relative humidity were found to be important in fluctuations of pod borer populations in Nigeria (Flood and Gilmour 2017).

For cacao diseases, a potential problem is that under climate change, pathogens currently considered of minor importance, and others not even found on cacao, might become hugely problematic. A good example of this is the pathogen *Lasiodiplodia theobromae*, which causes pod charcoal rot and tip dieback on cacao. This fungal pathogen appears to benefit greatly when cacao trees are stressed, especially under drought. Although currently considered a minor pathogen, in recent years severe outbreaks have been recorded (Mbenoun et al. 2008). Then there is the redistribution of cacao pathogens. With climate change, as cacao cultivation within countries shifts from less suitable to more suitable regions, pathogens are going to follow. This is likely to happen naturally, but here is a real danger that pathogens will be spread accidentally as farmers move to areas where they can grow cacao. Another threat is the effects on the cacao tree from the combined onslaught of environmental changes and disease. Trees already weakened by disease might succumb when also faced with drought, for example (Flood and Gilmour 2017).

Cacao Breeding and Climate Change

Looking to the future, new cacao varieties will need to be able to cope with heat and drought stress, as well as pests and pathogens. Researchers argue that the traits to enable cacao to deal with these stresses are probably already present in the cacao germplasm. After all, in the wild, cacao is found over a wide geographical area encompassing considerable variation in climate, and some of this valuable variation is already present in international cacao gene-banks

(Farrell et al. 2018). A great deal of the cacao grown today comes from material bred in South America and the cacao distributed from this region tends to possess low genetic diversity. But all is not lost, because cacao grown outside its center of origin in South America might be a valuable source of traits to enable the plant to cope with these stresses, having already proved itself capable of coping with drought in West Africa or high temperatures in Asia. Also, we should not, however, rule out cacao growing in the wild as sources of genetic diversity, especially since large areas of wild cacao are still to be explored. The problem here is to get to these areas before forests are destroyed (Farrell et al. 2018).

One way of increasing the genetic diversity of cacao in the short term is the use of seed gardens, which require having materials known to carry traits of interest. Here, different but compatible cacao plants are grown together to allow them to cross-pollinate. The seeds from these crosses can then be used for planting. Seed gardens have been used in Africa and Brazil and not only increase genetic diversity through cross-pollination, but also help to maintain some control over the consistency of planting material used by farmers (Farrell et al. 2018).

Some researchers argue that increasing the efficiency of cacao breeding would be best achieved by involving farmers in selecting and validating new varieties (Farrell et al. 2018). For crops such as cacao, landraces (locally adapted varieties) provide a lot of valuable genetic diversity, and on-farm evaluations have uncovered populations of cacao with a wide range of genetic traits. If farmers can be persuaded to participate in selecting such material, and data collection can be standardized, this could provide useful improvements in cacao germplasm adapted to different areas.

Crucial to future efforts to breed new cacao varieties is the material held in the two International Cocoa Genebanks, in Trinidad and Costa Rica. As important as these collections are, however, their funding for the long term is not secure. Over the past decade or so, efforts have been made to support their continued operation. In 2005, a proposal was made to create a network to optimize the conservation and utilization of cacao genetic resources worldwide for the benefit of breeders, researchers, and farmers. The result was CacaoNet (CacaoNet.org, n.d.), which was launched in 2006, and continues to support the two International Genebanks as well as the International Cocoa Quarantine Centre at the University of Reading in the UK. Also important are the various national cacao collections across the cacao-growing regions of the world. Unfortunately, these are underutilized and, as such, are at risk, especially since funding for their continued operation is insufficient and not guaranteed.

Producing cacao varieties that can cope with future climate change is a major task for the network of regional nonprofit organizations that collect and select cacao germplasm. This will require a long-term effort that will be difficult to implement if the network continues to be underfunded. What is needed is a coordinated, long-term approach, funded by both the public and private sectors (Farrell et al. 2018).

As with other perennial tree crops, breeding new cacao varieties is a long-term business. But it doesn't need to be, because the whole process can be speeded up using new genetic tools. I refer to gene editing, where specific, targeted changes to the genome can be made. It uses the bacterial approach to deal with a virus infection, and the way it works is ingenious. Basically, the bacteria set out to destroy the viral genome, and since the virus needs its genome to replicate, it's doomed. When a bacterium is attacked by a virus it hasn't encountered previously, it copies sections of the virus DNA and stores them in its own genome, producing a sort of genetic memory of the virus. The next time it encounters the virus, it uses an enzyme to chop up any DNA sequences matching the stored virus DNA. Bacteria have many stored sequences of virus DNA from previous virus attacks, and these are known as Clustered Regularly Interspaced Short Palindromic Repeats, mercifully known by their acronym, CRISPR. Rather like genetic scissors, CRISPR allows researchers to make precise changes at specific locations in the genome, making it possible to insert or replace specific genes or disrupt their function (Doudna and Charpentier 2014). This produces changes in the plant that are not unlike those found in naturally occurring plant populations (Zhang et al. 2018). As a result, the risks involved in gene editing are significantly lower than those associated with genetically modified crops. In fact, there would be no way of distinguishing between a gene-edit and a "naturally occurring mutation" (Zhang et al. 2018).

Gene editing has the potential to accelerate the breeding of new cacao varieties that are better equipped to deal with climate change, as well as diseases and pests, and indeed, research using CRISPR technology to improve cacao is already under way. For example, Mark Guiltinan and his colleagues at Penn State University in the USA are using gene editing tools to improve disease resistance in cacao. This work has resulted in plants exhibiting strong resistance to *Phytophthora tropicalis,* a pathogen of cacao and other crops (Guiltinan-Maximova Lab, n.d.; Fister et al. 2018). Gene editing research is also under way in the Innovative Genomics Institute at the University of California, Berkeley. The aim of this work, involving collaboration with chocolate manufacturer Mars Inc., is to make cacao more resilient to climate change (Innovative Genomics Institute 2018). Mars Inc. has also joined forces with scientists at the University of California, Davis, to develop a research facility that will focus on breeding improved cacao clones with, for example, resistance to diseases and pests (Boykin 2019).

. . .

It's a sad fact that cacao growers get less than 10% of the sale price of chocolate. Not only that, but most cacao growers have never actually tasted chocolate (many don't even know what the cacao beans they produce are used for) (Farrell et al. 2018). It stands to reason, therefore, that the costs associated with breeding new cacao varieties cannot be borne solely by cacao growers or their host countries. At present, funding for cacao breeding comes largely from multinational chocolate companies. Since the companies that benefit from cacao are largely located far away from the countries that produce it, perhaps the funding for breeding new cacao varieties should be linked to profits from trading and selling cacao (Farrell et al. 2018).

15

Chocolate

Luxury and Livelihood

To many of us across the world, chocolate is a readily available treat. We can walk into any supermarket and choose from a bewildering variety of chocolate sweets, most of which are easily affordable. We pick our favorite off the shelf and think only of how soon before we get to devour it. Some of us might actually scan the chocolates available before we make our choice, looking at what percentage of cocoa the chocolate contains, or the country (or countries) of origin, or whether it is Fairtrade or Rainforest Alliance certified. But how many are aware of the fact that 60% of the world's chocolate comes from two countries in West Africa (FAO n.d.), where it is produced by smallholders working just a few acres of land, and who are unlikely to have ever tasted chocolate? In fact, some 90% of the worldwide cacao crop is grown by smallholder farmers, on an estimated 4.5 million cacao farms (Make Chocolate Fair, n.d.). For these farmers, cacao is a cash crop, accounting for a large percentage of family income in many countries. Farmer incomes from cacao, however, are low, and although they produce the beans, their income represents a fraction of the price of a bar of chocolate (Ryan 2011). Most of the value of a bar of chocolate goes to the manufacturer, because the majority of the costs are in research and development, and in marketing (Nieburg 2014).

In West Africa, especially Côte d'Ivoire and Ghana, yield per tree is low, due to a combination of factors, including the age of the trees, high incidence of diseases and pests, and lack of nutrients in the soil (Wessel and Quist-Wessel 2015). But it doesn't matter where in the world the cacao is grown, the tree is susceptible to a range of diseases and pests that can destroy anywhere from 30% to the entire crop. Because of the limited availability of improved seeds and planting material, farmers harvest from old trees which produce low pod yields. Often, owners of small cocoa farms don't earn enough from the crop to buy fertilizer and chemicals, although efforts have been made to help farmers. In Côte d'Ivoire and Ghana, for example, the respective governments have launched schemes to provide farmers with improved plant material, fertilizers, and crop protection chemicals (Wessel and Quist-Wessel 2015). This might increase yields in the short term, but looking further into the future, climate change and the need to use more land for food production will provide cacao farmers with new challenges. In West Africa, climate change is projected to reduce the suitable area for cacao production, and coupled with more land being used to produce food, less land will be available for growing cacao. If West Africa is to remain at the top of the leader-board for cacao output, substantial changes are required.

A survey in 2001 for the Sustainable Tree Crop Programme (STCP), found that for the top 25% of West African cacao farms, average production costs were four times lower and yields almost four times greater than the bottom 25% of cocoa farms (Wessel and Quist-Wessel 2015). The STCP recommended selective replanting with improved material on larger cacao farms, although this would need to be made financially attractive to farmers, who traditionally prefer to plant cacao in new areas. For less efficient farmers, the STCP recommended that support be provided to allow them either to shift to alternative crops, or to use a low-input system, such as intercropping cacao with other economically valuable trees (Wessel and Quist-Wessel 2015).

Following the STCP's recommendations would mean reducing the area of extensive, low-yielding cacao and moving to a more intensive system of cultivating the crop. Cacao would need to be replanted on suitable soils, and farmers would have to invest in high-quality plant material as well as plant appropriate shade trees. Priority would also need to be given to controlling diseases and pests, which, left uncontrolled, would undo all the other good work undertaken by farmers. The problem is that this approach would require a complete change of mind-set for most cocoa farmers, who want to incur little cost in growing their cacao and would prefer to invest their income from the crop in the health and education of their children, rather than spend it on the cacao trees (Wessel and Quist-Wessel 2015). Ultimately, climate change is likely to dictate where cacao can best be grown, because, as we've already seen, not all regions of West Africa will continue to be suitable for growing cacao.

A great deal of tropical forest has been lost to cacao planting, with huge ecological consequences. Loss of forests is also of concern because of their importance for the climate, both locally and globally. Forest loss at a local level is associated with reduced rainfall and increased temperatures, while at a global level, it will contribute to greenhouse gas emissions. An investigation by the Guardian newspaper and the global campaign organization Mighty Earth, published in 2017, found that not only was cacao driving deforestation in West Africa, it seemed likely it was doing the same in other cacao-growing regions as well (MacLean 2017; Mighty Earth, n.d.). They found evidence of large-scale deforestation in Peru, Ecuador, Indonesia, and Cameroon and warned that if the industry didn't change, the huge loss of forests in Côte d'Ivoire and Ghana, and the consequences for the ecology of the affected areas and cacao cultivation, could also occur in these countries. Following the investigation, 23 of the world's largest chocolate companies, together with the governments of Côte d'Ivoire and Ghana, signed on to "no new deforestation in West Africa." Only two companies, Hershey and Olam, said they would commit to deforestation-free cacao globally (MacLean 2017; Mighty Earth, n.d.). Disappointingly, in the year since the pledges were made, tens of thousands of hectares of forest were cleared in Côte d'Ivoire and Ghana, and campaigners are concerned that the cacao industry in Africa is failing on its deforestation pledge (MacLean 2018).

To reduce further deforestation and ecological disturbance, planning is needed to determine areas of forest that could usefully be planted to cacao and areas that should be preserved. Also of ecological concern is the mass spraying of insecticides on cacao farms. This can affect, among other things, numbers of pollinating insects, with consequences for the cacao crop. It has been suggested that farmers be paid a premium for removing low-yielding cacao trees, thereby helping to reduce the areas treated with insecticide. Reducing the use of crop protection chemicals is highly desirable, but that requires the availability of alternative approaches to protecting the cacao crop from diseases and pests (Wessel and Quist-Wessel 2015). This should be research priority for the future.

There are clearly huge challenges facing cacao and the industry responsible for transforming it into chocolate. At the base of the chain that leads ultimately to the chocolate bar are the cacao farmers, several million of whom are responsible, in Côte d'Ivoire and Ghana, for producing 60% of the world's cacao. Here, as in some other cacao-growing countries, cultivating cacao is a family affair. In Côte d'Ivoire and Ghana, some of the cacao farmers are too poor not to have their children working on the farm. But child labor on cacao farms can include more than children of the family—it can also include children from neighboring areas or from countries who have left their homes in search of paid employment. Coming, for example, from drought-stricken areas where food is in short supply, their intention is to earn enough money to take some back home, to help out their struggling families. The reality, however, can be very different. Taken to farms with the promise of getting paid for their work, they find themselves working for no money, and with barely adequate food and shelter (Off 2006; Ryan 2011). There was an international outcry when the use of child labor on cacao farms in West Africa was discovered. Nearly 20 years later, the thorny issue of child labor won't go away.

The International Cocoa Initiative (ICI), a Swiss-based foundation that brings together the chocolate industry, farming communities, and national governments, started operating in Côte d'Ivoire and Ghana in 2007. Its aim is to safeguard child rights and help eliminate child labor. ICI set up the child labor monitoring and remediation system, which several large chocolate companies use in an attempt to prevent the use of child labor on cacao farms. Despite the use of this monitoring system, the ICI report for 2017 noted that nearly 15,000 cases of child labor were identified on cacao farms in Côte d'Ivoire and Ghana in 2012. Apparently, because the monitoring system depends on cacao farmers being honest about whether children work on their farms, and they fear being punished if they are identified as using child labor, a considerable number of cases go unidentified. Indeed, the 15,000 cases reported for 2012 are, according to the ICI, likely to be a huge underestimate (International Cocoa Initiative 2017).

Identifying cases of child labor is one thing, being able to do something about it is a different matter entirely. Poverty has a huge role to play. Perhaps the father of a family has died, leaving the mother to look after several children. Here, the eldest child might have to help. Should the family be punished? I would say not, so how should the issue of child labor be tackled? The ICI says the answer lies in better education, for both the parents and their children. For poor families this would be impossible without financial support, which might mean helping the mother to find paid employment. But what about work on the cacao farm? Who would do that if the mother is away and only the children are left? The ICI says that they would set up a group of young men in the community, who would undertake some of the hazardous tasks (e.g., pruning with machetes, or spraying chemicals without protection) that might otherwise be given to children. Any cases of child labor where there is clear evidence of abuse and exploitation are reported to the relevant authorities (International Cocoa Initiative 2017).

Some believe that certification through credible, independent standards bodies such as Fairtrade, UTZ, and Rainforest Alliance, is vital in attempting to eradicate child trafficking and child labor in the chocolate industry (Stop the traffik, n.d.). Certification is an important step for chocolate manufacturers toward sustainability in cacao production, and in 2015, an estimated 23% of the global cacao area was certified by one of the four major labels—UTZ, Rainforest Alliance (these two joined forces in early 2018), Fairtrade, and Organic (Nieburg 2018). An important component of these certification schemes is the price premium. With some labels, such as UTZ and Rainforest Alliance, cacao premiums are not fixed, and the farmer and buyer negotiate the premium. With other labels, such as Fairtrade, there is a guaranteed premium. So, what has been the impact of certification on incomes for cacao farmers, levels of poverty, and child labor? Not enough, because according to some, premiums paid to cacao farmers are not sufficient to take them out of poverty. It seems that a cacao farmer in Côte d'Ivoire, certified by UTZ, has a per capita income for his household of US$1.40 a day, which falls short of the World Bank's poverty line of US$1.90 a day (Nieburg 2018).

The Fourth World Cocoa conference was held in Berlin in the early summer of 2018. According to one of the declarations made during the conference, the cacao sector will not be sustainable if farmers are not able to earn a living income. It goes on to say that all stakeholders should develop and implement policies that enable cacao farmers to make a living income. It turns out that a living income for a cacao farmer in Côte d'Ivoire is US$6,133 a year for a household of eight people. This equates to US$2.10 per person per day (Nieburg 2018). Clearly, there is some way to go before cacao farmers in Côte d'Ivoire see such an increase in income.

One of the effects of low incomes for cacao growers is the effect it has on the younger generation of farmers. In West Africa, many young farmers don't want to grow cacao, which is understandable given the threats facing the crop, and the low income they can expect. The average age of a cacao farmer in Ghana is 52, which makes the reluctance of the younger generation to grow the crop worrying (Sethi 2017).

In an effort to tackle the issues of cacao sustainability, including sustainable livelihoods for cacao farmers, CocoaAction was launched in 2014. This is a voluntary strategy that aligns the leading chocolate companies, the governments of Côte d'Ivoire and Ghana, and important stakeholders on regional issues of cacao sustainability. Chocolate companies committed to CocoaAction include Mars Inc., Mondelēz International, Nestlé, The Hershey Company, and Barry Callebaut. Their vision is "to enable a sustainable and thriving cocoa sector where farmers prosper, cocoa-growing communities are empowered, human rights are respected, and the environment is conserved" (Sethi 2017). According to Rick Scobey, president of the World Cocoa Foundation, the body which convenes CocoaAction, their most important target is to raise cacao farmers above the poverty line of US$1.90 per day. After that, they want to ensure the farmers have enough income to afford school fees, crop inputs, housing, health, and safe water and sanitation (Nieburg 2018). Regarding cacao sustainability, CocoaAction aims to boost productivity. They claim that by increasing productivity on environmentally suitable land and boosting farmer income, it will be possible to reduce the pressure on farmers to encroach on new forests, "to produce more cacao on less land" (CocoaAction 2016). These are early days, since CocoaAction was only launched in 2014 and its strategy was put into practice in 2016.

Then, in 2019, Ghana and Côte d'Ivoire agreed to institute a per-tonne "living income differential" as part of a new payment structure designed to safeguard cacao farmers against falling prices (Coffee and Cocoa International 2019). The two countries decided to implement a US$400 per metric tonne differential in the 2020–2021 season. They will legislate that the minimum price a farmer can receive for cacao is US$1,820 per metric tonne, which is reckoned to cover the cost of production plus 13% for farmer income. In order to pay farmers US$1,820 and allow them to cover their own costs, Ghana and Côte d'Ivoire need to get a free on-board price (the price invoiced or quoted by a seller including all charges up to placing the goods on board a ship in port) of US$2,600. This would allow 70% to be passed on to farmers, with the remaining 30% going toward the costs incurred by Ghana and Côte d'Ivoire. Basically, the US$400 living income differential will be paid on top of the free on-board price all the time, even when the market is high (TCHO Chocolate 2019).

• • •

There is no doubt that the long-term sustainability of cacao is threatened by the issues dealt with in this book: diseases, pests, and climate change, as well as other factors, including deforestation, poverty, and child labor. It's easy to get caught up in the pessimism and panic invoked by headlines such as "CHOCCO-GEDDON—Experts fear chocolate will run out in THIRTY YEARS because cacao plants are dying in warmer climate" (Allen 2018), "The cocoa crisis: why the world's stash of chocolate is melting away" (Ford et al. 2014), and "Could chocolate actually go extinct in 40 years?" (Lee 2018). There is even the title of this book—*Chocolate Crisis: Climate Change and Other Threats to the Future of Cacao*. Although it is true that climate change, or the climate crisis as it is now being called, threatens cacao production across the cacao-growing regions of the world, cacao will not go extinct, and it will still be possible to grow the crop. Advances in plant breeding and biotechnology mean that we can speed up the breeding process, allowing the generation of new cacao varieties better able to cope with diseases, pests, and climate change. Coupled with programs such as the World Cocoa Foundation's Climate Smart Cocoa (World Cocoa Foundation, n.d.), which aims to increase industry engagement and investment in climate-smart agriculture, the future for cocoa and chocolate is not as bleak as the headlines proclaim.

As Kristy Leissle says in her book *Cocoa* (Leissle 2018), chocolate is a luxury good, and although we might crave it, we don't need it to survive. Nevertheless, millions of people look forward to eating some chocolate most days, and that is not likely to change anytime soon. It is my hope, and I'm sure the hope of those in the global chocolate manufacturing business, that it's not just the cacao tree that survives, but also the many people across the world who grow it.

Glossary

agroforestry: land use management system in which trees are grown among or around crops. Its benefits include increased biodiversity and reduced soil erosion.

Amelonado: a subspecies of Forastero cacao, it is the most extensively planted variety.

appressorium: a specialized cell typical of many fungal and Oomycete pathogens used to infect plant cells. It adheres tightly to the plant surface and from it a tiny peg is produced which enters the plant, typically using a combination of softening of the plant tissue and great pressure.

Aztec: a Mesoamerican culture that flourished in central Mexico in the Postclassic period from 1300 to 1521. The Aztecs comprised different ethnic groups of central Mexico, especially those who spoke the Nahuatl language. Aztec culture was organized into city-states, some of which formed alliances. The Aztec empire was established in 1427 as a confederation of three city-states: Tenochtitlan, Texcoco, and Tlacopan.

basidiospore: sexual spore produced by Basidiomycete fungi, which include mushrooms, rusts, and the cacao pathogens, witches' broom (*Moniliophthora perniciosa*) and frosty pod rot (*M. roreri*).

biotrophy: parasitic relationship in which the parasite feeds on its plant host without killing it.

biotype: a group of organisms with an identical genetic constitution.

black pod: disease of cacao caused by pathogens of the genus *Phytophthora*, mainly *P. palmivora* and *P. megakarya*. It can affect most parts of the cacao tree, although it is usually associated with the pod.

cacao: refers to the tree, *Theobroma cacao*, and the raw materials derived from it: pods and beans.

cacao swollen shoot virus (CSSV): virus that infects primarily cacao trees. It is transmitted by mealybugs and is endemic in Togo, Ghana, and Nigeria.

Cacao Nacional: a variety of Amelonado cacao planted in Ecuador.

caffeine: an alkaloid found in cacao and chocolate. It is also found in coffee and tea and acts as a stimulant of the central nervous system.

cauliflory: literally translated as "stem-flower," the term refers to flowers and inflorescences that develop directly from the trunks, limbs, and main branches of woody plants.

cherelle: a small cacao pod before it matures.

chocolate: obtained once cocoa powder is mixed with milk, sugar, and cocoa butter.

chromosome: in cells, DNA is packaged into threadlike structures called chromosomes. In a chromosome, the DNA is tightly coiled many times around proteins called histones that support its structure.

chupon: the main stem of a young cacao tree.

clone: a cell or an organism that is genetically identical to the original cell or organism from which it is derived.

cocoa: technically, the product produced once the beans have been roasted and ground. It is also used in some parts of the world—Africa, southeast Asia, and Britain, for example, to refer to the tree, its pods and beans, as well as the powder produced by roasting and grinding the beans.

cocoa pod borer: larva of the minute moth, *Conopomorpha cramerella*. It is the most damaging insect pest of cacao in southeast Asia.

Comum: a variety of Amelonado cacao planted in Brazil.

conching: a long process of intense mixing, stirring, and aerating of heated liquid chocolate to eliminate unwanted acidity and bitterness in the final product. The conching process was developed by Rodolphe Lindt in 1879.

Criollo: in Spanish, "of local origin." The beans produced by this variety of cacao are of very high quality, very aromatic, yet lacking in bitterness. It is becoming rare and expensive and represents less than 5% of the world's chocolate production.

detritivore: an animal that feeds on dead organic matter, especially dead plant material.

Dutching: the process in which cocoa mass or cacao nibs are treated with an alkali, such as potassium carbonate. Its original purpose was to increase the dispersion properties of the cocoa powder when mixed with milk, although it is also used to intensify color. It reduces the acidity and astringency of the cacao beans and cocoa mass and can help to improve and intensify the aromatic characteristics associated with cocoa, although it can reduce the polyphenol content of cocoa powder by as much as 80%.

El Niño: a climate cycle in the Pacific Ocean with a global impact on the Earth's weather patterns. It begins when warm water in the western Pacific shifts eastward along the equator toward the coast of South America. Usually, the warm water sits near Indonesia and the Philippines, but during an El Niño the warm waters sit off northwestern South America. El Niño phases occur about every four years, although cycles have lasted between two and seven years.

El Niño–Southern Oscillation: an irregularly periodic variation in winds and surface sea temperatures over the tropical eastern Pacific Ocean. The warming phase is called El Niño and the cooling phase is known as La Niña. The El Niño–Southern Oscillation affects the climate of much of the tropics and subtropics.

endophyte: often a bacterium or fungus that lives within a plant without causing disease.

epidemiology: the study of how disease develops in populations.

Forastero: a cacao variety with strong, earthy flavors, representing 80% of the world's cacao production. Forastero means "stranger" or "outsider" in Spanish.

flavanols: a group of flavonoid polyphenols, including catechins and procyanidins, found in high concentrations in cocoa, grapes, red wine, and tea. They occur naturally in plants, where many play a protective role against attack by pathogens and pests, and against other stresses.

Florentine Codex: a sixteenth-century ethnographic research study undertaken in Mesoamerica by the Franciscan friar Bernardino de Sahagún. The best-preserved manuscript is held in the Laurentian library in Florence.

frosty pod rot: disease of cacao caused by the fungus *Moniliophthora roreri*. It originated in western Colombia/Ecuador, but has spread to Peru,

Venezuela, and Bolivia, and into Central America as far as Mexico. Africa, Asia, and some of the Caribbean are still free of the fungus.

gene: comprising DNA, genes are the basic physical and functional unit of heredity. Some genes act as instructions to make proteins. The cacao tree has approximately 35,000 genes, while the human genome comprises between 20,000 and 25,000 genes.

genetic fingerprinting: a method for detecting variable sequences within DNA to distinguish individuals from one another. It is also known as DNA fingerprinting or DNA profiling.

genome: the complete set of an organism's DNA, including all its genes.

genotype: in a broad sense, the term refers to the genetic makeup of an organism. Genotype can also be used in a narrow sense to refer to the variant forms of a gene (known as alleles) carried by an organism.

global warming: long-term increase in average global temperature of the Earth's climate and a major aspect of climate change. Although the terms global warming and climate change tend to be used interchangeably, climate change refers to global warming and its effects, which include changes in precipitation, for example, which differ across the world's regions.

greenhouse effect: the effect of atmospheric gases such as carbon dioxide absorbing energy from the sun and Earth and trapping it near the Earth's surface. This warms the Earth, making it hospitable for life. Too little of the greenhouse effect would cool down the Earth, while too much would warm it up, in both cases making life on the planet difficult.

greenhouse gas: a gas that allows the sun's rays to pass through and warm the Earth, but prevents the warmth from escaping back into space, thereby contributing to the greenhouse effect. Greenhouse gases include carbon dioxide, methane, and water vapor.

haustorium: feeding structure produced by certain fungal and Oomycete pathogens. It enters the plant cell but never ruptures the cell's plasma membrane.

hemibiotroph: a pathogen that has an initial biotrophic phase within the plant, when it causes minimal damage, but then switches to a necrotrophic phase when it kills the cells and lives off the dead material.

infrared radiation: a type of radiant energy invisible to human eyes but which we can feel as heat. The most obvious source of infrared radiation is the sun, although all objects in the universe emit some level of it. It is electromagnetic radiation with wavelengths longer than those of visible light, extending from 700 nm to 1 mm. The balance between absorbed and emitted infrared radiation has a critical effect on Earth's climate.

instar: a phase between two periods of molting during the development of an insect larva.

K'iche': Mayan people living in the midwestern highlands of Guatemala. They had an advanced civilization in pre-Columbian times, with a high level of political and social organization. Today, the K'iche' number around 1 million, and their language is one of the largest Mayan linguistic groups. The K'iche' reached the peak of their power and influence during the Mayan Postclassic period (c. 950–1539). K'iche' means "many trees," and the Nahuatl translation, "Cuauhtēmallān," is the origin of the word Guatemala. The previous Spanish spelling of K'iche' was Quiché.

Lacandon: one of the Mayan peoples living in the jungles of Chiapas in Mexico, near its southern border with Guatemala. They are one of the most isolated and culturally conservative of Mexico's native peoples. Although they had almost died out by the early 1940s, their population has grown significantly since then and currently stands at around 300.

mass: formed when roasted cacao nibs are finely ground to produce a soft, silky liquid.

Maya: a Mesoamerican civilization that encompassed southern Mexico, all of Guatemala and Belize, and the western portions of Honduras and El Salvador. Maya is a modern, overarching term referring to the peoples of the region, although they have never used the term themselves, since there has been no sense of identity or political unity among the different populations making up the Maya. The Mayan culture was noted for its art, architecture, mathematics, calendar, and astronomical system, and had the most sophisticated and highly developed writing system in the Americas in pre-Columbian times.

Mesoamerica: historical region and cultural area in North America, extending from central Mexico through to northern Costa Rica, encompassing Belize, Guatemala, El Salvador, Honduras, and Nicaragua. Prior to the colonization of the Americas by the Spanish, pre-Columbian societies flourished in this region.

microbiome: a plant microbiome is the community of microbes that typically interact with the plant. It includes those microbes that interact with roots and leaves, which commonly reside in the rhizosphere and phyllosphere, respectively, and those microbes that reside within the plant, known as endophytes.

mirid: insect belonging to the insect family Miridae, comprising more than 10,000 species. Plant mirids feed by inserting their needlelike mouthparts into the host and sucking out the sap. Not only is their saliva toxic to plants, but while feeding, mirids can also transmit viruses.

Nahuatl: a group of languages of the Uto-Aztecan language family. It has been spoken in central Mexico since at least the seventh century and was the language of the Aztecs. It is still spoken by some 1.7 million Nahua peoples.

necrotrophy: parasitic relationship in which the parasite kills the plant cells and tissues prior to feeding on the dead matter.

nibs: technically, these are the cotyledons of the cacao bean. Nibs are left following the winnowing process, which removes the outer shell of the bean.

Olmecs: the first civilization of the Americas, they flourished from 1500 to 400 BC. The Olmec civilization was centered in the humid lowlands of the Mexican Gulf coast. They spoke an ancestral form of the Mixe-Zoquean family of languages.

Oomycetes: also known as "water molds." They were once regarded as lower fungi, on account of their filamentous growth habit, nutrition by absorption, and reproduction via spores. It is now clear, however, that oomycetes are unrelated to true fungi; the latter appear more closely related to animals, and oomycetes are more closely related to algae and green plants. Oomycetes include some of the world's most devastating plant pathogens.

photosynthesis: process by which plants use carbon dioxide and sunlight to produce sugars. It is carried out in organelles in plant cells known as chloroplasts.

polyphenols: a group of chemicals that occur naturally in plants. Polyphenols can be categorized into four groups: flavonoids, phenolic acids, stilbenes, and lignans. Health benefits associated with polyphenols may be related to their role as antioxidants, which can combat cell damage. In plants, many polyphenols play a role in protecting plants against various stresses, such as attack by pathogens and pests.

resistance: the ability of a plant to limit the growth and development of a pathogen. Resistance can be determined by one or a few genes, in which case it is called monogenic resistance, or it can be determined by many genes, in which case it is known as polygenic resistance.

saprotrophy: the ability to obtain nourishment from dead or decaying organic matter.

sporangium (pl. **sporangia**): a structure that encloses reproductive spores; in the case of Oomycete pathogens such as black pod, the spores are asexual.

stomata: pores in the epidermis of leaves, stems, and other plant organs that facilitate gas exchange, primarily the entry of carbon dioxide for photosynthesis and the loss of water vapor in transpiration.

theobromine: the main alkaloid found in cacao and chocolate. Like caffeine, it stimulates the central nervous system, but to a lesser extent. The word is derived from the genus name for the cacao tree, *Theobroma cacao*, and comprises two parts, both with Greek roots, *theo* (gods) and *broma* (food).

tolerance: the ability of a plant to endure pathogen attack and colonization without serious effects on its own growth and development.

Trinitario: this cacao variety is the result of cross-fertilization between Forastero cacao and Criollo cacao in Trinidad. The Forastero seeds were taken to Trinidad following nearly complete destruction of Trinidad's Criollo plantations in 1727 by what is now thought to have been a disease, rather than a hurricane as originally thought. Trinitario beans form about 12% of the world's cacao production.

vascular streak dieback: disease of cacao caused by the basidiomycete fungus *Ceratobasidium theobromae*. It is one of the most important diseases in the southeast Asian/Melanesian region.

virus: a parasite comprising a central core of DNA or RNA surrounded by a protein coat. It cannot reproduce by itself; once it infects a host, it hijacks its cell machinery, forcing the host cells to produce more virus particles.

witches' broom: fungal disease of cacao caused by *Moniliophthora perniciosa*. It is restricted to South America, parts of Central America, and the Caribbean but is absent from Africa and Asia, where 85% of the world's cacao is grown.

zoospore: a spore produced by some fungi and Oomycetes capable of swimming using a flagellum.

References

Abdulai, I, Vaast, P, Hoffmann, MP, Asare, R, Jassogne, L, Van Asten, P, Rötter, RP, and Graefe, S. 2017. Cocoa agroforestry is less resilient to sub-optimal and extreme climate than cocoa in full sun. *Global Change Biology* 24: 273–286.

Abdulai, I, Jassogne, L, Graefe, S, Asare, R, Van Asten, P, Laderach, P, and Vaast, P. 2018. Characterization of cocoa production, income diversification, and shade tree management along a climate gradient in Ghana. *PLOS ONE* 13(4): e0195777. https://doi.org/10.1371/journal.pone.0195777.

Abou Rajab, Y, Leuschner, C, Barus, H, Tjoa, A, and Hertel, D. 2016. Cacao cultivation under diverse shade cover allows high carbon storage and sequestration without yield losses. *PLOS ONE* 11(2): 1–22.

Addo-Fordjoura, P, Gyimah Gyamfib, H, Fei-Baffoec, B, and Akrofid, AY. 2013. Impact of copper-based fungicide application on copper contamination of cocoa plants and soils in the Ahafo Ano North District, Ashanti Region, Ghana. *Ecology, Environment and Conservation Paper* 19(2): 303–310.

Adjaloo, M, Banful, B, and Oduro, W. 2013. Evaluation of breeding substrates for cocoa pollinator, *Forcipomyia* spp. and subsequent implications for yield in a tropical cocoa production system. *American Journal of Plant Sciences* 4: 204–211.

Adu-Acheampong, R, Jiggins J, van Huis, A, Cudjoe, AR, Johnson, V, Sakyi-Dawson, O, Ofori-Frimpong, K, Osei-Fosu, P, Tei-Quartey, E, Jonfia-Essien, W, Owusu-Manu, M, and others. 2014. The cocoa mirid (Hemiptera: Miridae) problem: evidence to support new recommendations on the timing of insecticide application on cocoa in Ghana. *International Journal of Tropical Insect Science* 34: 58–71.

Agriculture Department of Indonesia. 2008. Tree crop statistics of Indonesia 2007–2009, Cocoa. Secretariat of Directorate General of Estates, Agriculture Department of Indonesia.

Agrios, GN. 2005. *Plant Pathology* (5th edition). Burlington, MA, USA: Elsevier Academic Press.

Ahenkorah, Y, Akrofi, GS, and Adri, AK. 1974. The end of the first cocoa shade and manurial experiment at the Cocoa Research Institute of Ghana. *Journal of Horticultural Science* 49: 43–51.

Ahenkorah, Y, Akrofi, GS, and Adri, AK. 1987. Twenty years' results from a shade and fertiliser trial on Amazon cocoa (*Theobroma cacao*) in Ghana. *Experimental Agriculture* 23(1): 31–39.

Aikpokpodion, PE, Lajide, L, and Aiyesanmi, AF. 2010. Heavy metals contamination in fungicide treated cocoa plantations in Cross River State, Nigeria. *American-Eurasian Journal of Agricultural & Environmental Science* 8(3): 268–274.

Aime, MC, and Phillips-Mora, W. 2005. The causal agents of witches' broom and frosty pod rot of cacao (chocolate, *Theobroma cacao*) form a new lineage of Marasmiaceae. *Mycologia* 97: 1012–1022.

Ali, SS, Amonko-Attah, I, Bailey, RA, Strem, MD, Schmidt, M, Akrofi, AY, Surujdeo-Maharaj, S, Kolawole, OO, Begonde, BAD, ten Hoopen, GM, Goss, E, and others. 2016. PCR-based identification of cacao black pod causal agents and identification of biological factors possibly contributing to *Phytophthora megakarya's* field dominance in West Africa. *Plant Pathology* 65: 1095–1108.

Ali, SS, Shao, J, Lary, DJ, Kronmiller, BA, Shen, D, Strem, MD, Amonko-Attah, I, Akrofi, AY, Begoude, BAD, ten Hoopen, GM, Coulibily, K, and others. 2017. *Phytophthora megakarya* and *Phytophthora palmivora*, closely related causal agents of cacao black pod rot, underwent increases in genome sizes and gene numbers by different mechanisms. *Genome Biology and Evolution* 9: 536–557.

Allen, F. 2018. "CHOCCO-GEDDON—Experts fear chocolate will run out in THIRTY YEARS because cacao plants are dying in warmer climate." *The Sun*, January 1, 2018. https://www.thesun.co.uk/tech/5247207/experts-fear-chocolate-will-run-out-in-thirty-years-because-cacao-plants-are-dying-in-warmer-climate/.

Almeida, AAF, and Valle, RR. 2007. Ecophysiology of the cacao tree. *Brazilian Journal of Plant Physiology* 19: 425–448. http://dx.doi.org/10.1590/S1677-04202007000400011.

Almeida, DSM, Gramacho, KP, Cardoso, THS, Micheli, F, Alvim, FC, and Pirovani, CP. 2017. Cacao phylloplane: the first battlefield against *Moniliophthora perniciosa*, which causes witches' broom disease. *Phytopathology* 107: 864–871.

Alvim, P de T. 1977. "Cacao." In *Ecophysiology of tropical crops*, edited by P de T Alvim and TT Kozlowski, 591–616. New York: Academic Press.

Ambele, FC, Bisseleua Daghela, HB, Babalola, OO, and Ekesi, S. 2018. Soil-dwelling insect pests of tree crops in Sub-Saharan Africa, problems and management strategies—a review. *Journal of Applied Entomology* 142: 539–552.

Ameyaw, LK, Ettl, GJ, Leissle, K, and Anim-Kwapong, GJ. 2018. Cocoa and climate change: insights from smallholder cocoa producers in Ghana regarding challenges in implementing climate change mitigation strategies. *Forests* 9: 742. Doi:10.3390/f9120742.

Ameyaw, GA. 2019. Management of the cacao swollen shoot virus (CSSV) menace in Ghana: the past, present and the future. *IntechOpen* DOI: http://dx.doi.org/10.5772/intechopen.87009.

Ampofo, ST. 1997. The current cocoa swollen shoot virus disease situation in Ghana. *First International Cocoa Pests and Diseases Seminar,* Accra, Ghana, November 6–10, 1995, 175–178.

Andebrhan, T, Figueira, A, Yamada, AA, Cascardo, J, and Furtek, DB. 1999. Molecular fingerprinting suggests two primary outbreaks of witches' broom disease (*Crinipellis perniciosa*) of *Theobroma cacao* in Bahia. *European Journal of Plant Pathology* 105: 167–175.

Andres, C, Blaser, WJ, Dzahini-Obiatey, HK, Ameyaw, GA, Domfeh, OK, Awiagah, MA, Gattinger, A, Schneider, M, Offei, SK, Six, J. 2018. Agroforestry systems can mitigate the severity of cocoa swollen shoot virus disease. *Agriculture, Ecosystems and Environment* 252: 83–92.

Aneja, M, and Gianfagna, T. 2001. Induction and accumulation of caffeine in young, actively growing leaves of cocoa (*Theobroma cacao* L.) by wounding or infection with *Crinipellis perniciosa*. *Physiological and Molecular Plant Pathology* 59: 13–16.

Anikwe, JC, Omoloye, AA, Aikpokpodion, PO, Okelana, FA, and Eskes, AB. 2009. Evaluation of resistance in selected cocoa genotypes to the brown cocoa mirid, *Sahlbergella singularis* Haglund in Nigeria. *Crop Protection* 28: 350–355.

Apothecaryads. 2014. "What the apothecary ordered." https://apothecaryads.tumblr.com/post/96162203734/robert-gibson-and-sons-chocolate-worm-cakes.

Aprotosoaie, AC, Luca, SV, and Miron, A. 2015. Flavor chemistry of cocoa and cocoa products. *Comprehensive Reviews in Food Science and Food Safety* 15: 73–91.

Aragundi, J. 1974. *Evaluación de rendimientos e incidencia de enfermedades del cacao, en varias zonas ecológicas del Litoral Ecuatoriano.* Thesis, Ingeneiro Agrónomo, Universidade de Guayaquil, Ecuador. Quoted in Evans, HC. 2016. "Frosty pod rot (*Moniliophthora roreri*)." In *Cacao diseases: a history of old enemies and new encounters,* edited by BA Bailey and LW Meinhardt, 137–177. Switzerland: Springer International.

Argout, X, Salse, J, Aury, JM, Guiltinan, MJ, Droc, G, Gouzy, J, Allegre, M, Chaparro, C, Legavre, T, Maximova, SN, Abrouk, M, and others. 2011. The genome of *Theobroma cacao. Nature Genetics* 43(2): 101–109.

Arnold, AE, and Herre, EA. 2003. Canopy cover and leaf age affect colonization by tropical fungal endophytes: ecological pattern and process in *Theobroma cacao* (Malvaceae). *Mycologia* 95: 388–398.

Arnold, AE, Mejía, LC, Kyllo, D, Rojas, EI, Robbins, N, and Herre, EA. 2003. Fungal endophytes limit pathogen damage in a tropical tree. *Proceedings of the National Academy of Sciences USA* 100: 15649–15654.

Arnold, C. 2011. The sweet smell of chocolate: sweat, cabbage and beef. *Scientific American* October 31, 2011. http://www.scientificamericam.com/article/sensomics-chocolate-smell/.

Asare, R, and Sonii, D. 2010. *Planting, replanting and tree diversification in cocoa systems. Learning about sustainable cocoa production: a guide for participatory farmer training.* Manual No 2, Environment. Copenhagen: Development and Environment Series 13–2010, Forest and Landscape Denmark, 2010.

Asogwa, EU, Ndubuaku, TCN, Ugwu, JA, and Awe, OO. 2010. Prospects of botanical pesticides from neem, *Azadirachta indica* for routine protection of cocoa farms against the brown cocoa mirid—*Sahlbergella singularis* in Nigeria. *Journal of Medicinal Plants Research* 4: 1–6.

Attignon, SE, Lachat, T, Sinsin, B, Nagel, P, and Peveling, R. 2005. Termite assemblages in a West African semi-deciduous forest and teak plantations. *Agriculture, Ecosystems and Environment* 110: 318–326.

Avila-Lovera, E, Coronel, I, Jaimez, R, Urich, R, Pereyra, G, Araque O, Chacón, I, and Tezara, W. 2016. Ecophysiological traits of adult trees of Criollo cocoa cultivars (*Theobroma cacao* L.) from a germplasm bank in Venezuela. *Experimental Agriculture* 52: 137–153.

Awudzi, GK, Asamoah, M, Owusu-Ansah, F, Hadley, P, Hatcher, PE, and Daymond, AJ. 2016. Knowledge and perception of Ghanaian cocoa farmers on mirid control and their willingness to use forecasting systems. *International Journal of Tropical Insect Science* 36: 22–31.

Awudzi, GK, Cudjoe, AR, Hadley, P, Hatcher, PE, and Daymond, AJ. 2017. Optimizing mirid control on farms through complementary monitoring systems. *Journal of Applied Entomology* 141: 247–255.

Ayuk, ET, Duguma, B, Franzel, S, Kengue, J, Mollet, M, Tiki-Manga, T, and Zenkeng, P. 1999. Use, management and economic potential of *Irvingia gabonensis* in the humid lowlands of Cameroon. *Forest Ecology and Management* 113: 1–9.

Azhar, I. 1990. Studies on sclerotic layer hardness of cocoa pods. *MARDI Research Journal* 18: 63–69.

Babin, R, ten Hoopen, GM, Cllas, C, Enjairic, F, Yede, Gendre, P, and Lumaret, J-P. 2010. Impact of shade on the spatial distribution of *Sahlbergella singularis* in traditional cacao agroforests. *Agricultural and Forest Entomology* 12: 69–79.

Babin, R, Anikwe, JC, Dibog, L, Lumaret, J-P. 2011. Effects of cocoa tree phenology and canopy microclimate on the performance of the mirid bug *Sahlbergella singularis*. *Entomologia Experimentalis et Applicata* 141: 25–34.

Baker, RED, and Holliday, P. 1957. Witches' broom disease of cacao (*Marasmius perniciosus* Stahel). *Phytopathological Papers* 2: 1–42.

Bailey, BA, Crozier, J, Sicher, RC, Strem, MD, Melnick, R, Carazzolle, MF, Costa, GGL, Pereira, GAG, Zhang, D, and Maximova, S. 2013. Dynamic changes in pod and fungal physiology associated with the shift from biotrophy to necrotrophy during the infection of *Theobroma cacao* by *Moniliophthora roreri*. *Physiological and Molecular Plant Pathology* 81: 84–96.

Bailey, BA, Ali, SS, Akrofi, AY, Meinhardt, LW. 2016. "*Phytophthora megakarya*, a causal agent of black pod rot in Africa." In *Cacao diseases: a history of old enemies and new encounters*, edited by BA Bailey and LW Meinhardt, 267–303. Switzerland: Springer International.

Bailey, BA, Evans, HC, Phillips-Mora, W, Ali, SS, and Meinhardt, LW. 2018. *Moniliophthora roreri*, causal agent of cacao frosty pod rot. *Molecular Plant Pathology* 19: 1580–1594.

Bandeira, AG, Vasconcellos, A, Silva, MP, Constantino, R. 2003. Effects of habitat disturbance on the termite fauna in a highland humid forest in the Caatinga domain, Brazil. *Sociobiology* 42: 117–128.

Barreto, MA, Santos, JCS, Corrêa, RX, Luz, EDMN, Marelli, J, and Souza, AP. 2015. Detection of genetic resistance to cocoa black pod disease caused by three *Phytophthora* species. *Euphytica* 206: 677–687.

Barsics, F, Delory, BM, Delaplace, P, Francis, F, Fauconnier, ML, Haubruge, É, and Verheggen, FJ. 2016. Foraging wireworms are attracted to root-produced volatile aldehydes. *Journal of Pest Science* 90: 69–76.

Bartley, BGD. 1986. *Cocoa. Theobroma cacao.* FAO Plant Production and Protection paper, 70. Rome: FAO.

Beerling, D. 2007. *The Emerald Planet: how plants changed Earth's history.* Oxford, UK: Oxford University Press.

Bekele, FL. 2004. "The history of cocoa production in Trinidad and Tobago." In *Revitalisation of the Trinidad and Tobago Cocoa Industry,* edited by LA Wilson, Proceedings of the APASTT Seminar-Exhibition, September 20, 2003, St. Augustine, Trinidad. Trinidad and Tobago: Association of Professional Agricultural Scientists of Trinidad and Tobago.

Bignell, DE, Roisin, Y, and Lo, N. 2010. *Biology of Termites: a Modern Synthesis* (1st ed.). Dordrecht, The Netherlands: Springer. ISBN 978–90–481–3977–4.

Bisseleua, DHB, Missoup, AD, and Vidal, S. 2009. Biodiversity conservation, ecosystem functioning, and economic incentives under cacao agroforestry intensification. *Conservation Biology* 22: 1176–1184.

Bisseleua, HD. 2019. "Early detection can eradicate cocoa disease in West Africa." *World Cocoa Foundation blog.* Accessed August 2, 2019. https://www.worldcocoafoundation. org/blog/early-detection-can-eradicate-cocoa-disease-in-west-africa/.

Borg, A, and Siegel, A. 2009. "Early works on chocolate: a checklist" In *Chocolate: History, Culture and Heritage,* edited by LE Grivetti and H-Y Shapiro, 929–942. NJ, USA: John Wiley & Sons.

Bos, MM, Steffan-Dewenter, I, and Tscharntke, T. 2007. Shade tree management affects fruit abortion, insect pests and pathogens of cacao. *Agriculture, Ecosystems and Environment* 120: 201–205.

Boudjeko, T, Djocgoue, PF, Nankeu, JD, Mbouobda, HD, Omokolo, DN, and El Hadrami, I. 2007. Luteolin derivatives and heritability of resistance to *Phytophthora megakarya* in *Theobroma cacao. Australasian Plant Pathology* 36: 56–61.

Boykin, S. 2019. "Mars Inc. collaborates with UC Davis on new research facility." *Sacramento Business Journal,* March 19, 2019. https://www.bizjournals.com/sacramento/news/2019/03/19/mars-inc-collaborates-with-uc-davis-on-new.html.

Brasier, CM, and Griffin, MJ. 1979. Taxonomy of *Phytophthora palmivora* on cocoa. *Transactions of the British Mycological Society* 72: 111–143.

Brown, O, and Crawford, A. 2008. *Assessing the security implications of climate change for West Africa. Country case studies of Ghana and Burkina Faso.* Winnipeg, Manitoba, Canada: International Institute for Sustainable Development.

Brunt, AA. 1975. The effects of cocoa swollen shoot virus on the growth and yield of Amelonado and Amazon cocoa (*Theobroma cacao*) in Ghana. *Annals of Applied Biology* 80: 169–180.

Buitrago-Lopez, A, Sanderson, J, Johnson, L, Warnakula, S, Wood, A, Di Angelantonio,

E, and Franco, OH. 2011. Chocolate consumption and cardiometabolic disorders: systematic review and meta-analysis. *British Medical Journal* 343: d4488.

Bunn, C, Talsma, T, Läderach, P, and Castro, F. 2017. "Climate change impacts on Indonesian cocoa areas." Report for CIAT/Mondelēz/Cocoa Life. https://www.cocoalife.org/progress/future-of-cocoa-in-indonesia-a-study-of-climate-change

Cabrera, OG, Molano, EPL, José, J, Álvarez, JC, and Pereira, GAG. 2016. "*Ceratocystis* wilt pathogens: history and biology—highlighting C. *cacaofunesta* the causal agent of wilt disease of cacao." In *Cacao diseases: a history of old enemies and new encounters*, edited by BA Bailey and LW Meinhardt, 383–428. Switzerland: Springer International.

CacaoNet.org. n.d. "Global network for cacao genetic resources." Accessed November 28, 2018. http://cacaonet.org/.

Cadbury, n.d. Factsheet: "Chocolate manufacturing." http://www.cadburyworld.co.uk/schoolandgroups/~/media/

Candy Industry. 2017. "Global state of the confectionery industry: shattering stereotypes." June 15, 2017. Accessed January 29, 2018. http://www.candyindustry.com/articles/87756-global-state-of-the-confectionery-industry-shattering-stereotypes.

Cassano, CR, Silva, RM, Mariano-Neto, E, Schroth, G, and Faria, D. 2016. Bat and bird exclusion but not shade cover influence arthropod abundance and cocoa leaf consumption in agroforestry landscape in northeast Brazil. *Agriculture, Ecosystems and Environment* 232: 247–253.

Chaves, FC, and Gianfagna, TJ. 2007. Cacao leaf procyanidins increase locally and systemically in response to infection by *Moniliophthora perniciosa* basidiospores. *Physiological and Molecular Plant Pathology* 70: 174–179.

Chetschik, I, Kneubühl, M, Chatelain, K, Schlüter, A, Bernath, K, and Hühn, T. 2018. Investigations on the aroma of cocoa pulp (*Theobroma cacao* L.) and its influence on the odor of fermented cocoa beans. *Journal of Agricultural and Food Chemistry* 66: 2467–2472.

Chingandu, N, Kouakou, K, Aka, R, Ameyaw, G, Gutierrez, OA, Herrmann, H-W, and Brown, JK. 2017. The proposed new species, cacao red vein virus, and three previously recognized badnavirus species are associated with cacao swollen shoot disease. *Virology Journal* 14: 199. DOI 10.1186/s12985-017-0866-6.

Christian, N, Herre, EA, Mejía, LC, and Clay, K. 2017. Exposure to the leaf litter microbiome of healthy adults protects seedlings from pathogen damage. *Proceedings of the Royal Society B* 284: 20170641. http://dx.doi.org/10.1098/rspb.2017.0641.

Cleveland, CJ, Betke, M, Federico, P, Frank, JD, Hallam, TG, Horn, J, López, Jr, JD, McCracken, GF, Medellín, RA, Moreno-Valdez, A, and Sansone, CG. 2006. Economic value of the pest control service provided by Brazilian free-tailed bats in south-central Texas. *Frontiers in Ecology and the Environment* 4: 238–243.

Clough, Y, Faust, H, and Tscharntke, T. 2009. Cacao boom, cacao bust: endangered sustainability and opportunities for biodiversity conservation. *Conservation Letters* 2: 197–205.

CocoaAction. 2016. "Learning as We Grow: Putting CocoaAction into Practice." Annual Report 2016. https://www.worldcocoafoundation.org/about-wcf/cocoaaction/.

Cocoalife.org. n.d. "Combatting deforestation and building climate change resilience."

Accessed November 28, 2018. https://www.cocoalife.org/the-program/climate-change.

Coe, SD, and Coe, MD. 2013. *The true history of chocolate*. Third Edition. London: Thames & Hudson.

Coffee and Cocoa International. 2019. "Cocoa price floor falls apart—Côte d'Ivoire and Ghana agree 'living income differential.'" Accessed October 24, 2019. https://www.coffeeandcocoa.net/2019/07/08/cocoa-price-floor-falls-apart-cote-divoire-and-ghana-agree-living-income-differential/.

Cornwell, PB. 1958. Movements of the vectors of virus diseases of cacao in Ghana. I. Canopy movement in and between trees. *Bulletin of Entomological Research* 49: 613–630.

Cuatrecasas, J. 1964. Cacao and its allies: a taxonomic revision of the genus *Theobroma*. *Contributions from the U.S. National Herbarium* 35: 379–614.

Cubillos, G. 2017. *Frosty pod and pod borer: enemies that damage cocoa crops in Colombia*. Saarbrüken, Germany: Lambert Academic.

Curry, GN, Koczberski, G, Lummani, J, Nailina, R, Peter, E, McNally, G, and Kuaimba, O. 2015. A bridge too far? The influence of socio-cultural values on the adaptation responses of smallholders to a devastating pest outbreak in cocoa. *Global Environmental Change* 35: 1–11.

Dade, HA. 1927. Economic significance of cacao pod disease and factors determining their incidence and control. *Bulletin of the Department of Agriculture, Gold Coast* 6: 7–59.

Dawes, TZ. 2010. Impacts of habitat disturbance and soil water storage in a tropical Australian savannah. *Pedobiologia* 53: 241–246.

Day, RK. 1989. Effect of cocoa pod borer, *Conopomorpha cramerella*, on cocoa yield and quality in Sabah, Malaysia. *Crop Protection* 8: 332–339.

de Albuquerque, PSB, Silva, SDVM, Luz, EDMN, Pires, JL, Vieira, AMC, Demétrio, CGB, Pascholatti, SF, and Figueira, A. 2010. Novel sources of witches' broom resistance (causal agent (*Moniliophthora perniciosa*) from natural populations of *Theobroma cacao* from the Brazilian Amazon. *Euphytica* 172: 125–138.

Decazy, B, and Essono, B. 1979. Tests de contrôle d'infestation et traitements anti-mirides. *Café Cacao Thé* 23: 35–42.

Deheuvels, O, Avelino, J, Somarriba, E, and Malezieux, E. 2012. Vegetation structure and productivity in cocoa-based agroforestry systems in Talamanca, Costa Rica. *Agriculture Ecosystems & Environment* 149: 181–188.

de la Cruz, MT, García, CO, Ortiz, OT, Aguilera, AM, and Díaz, CN. 2011. Temporal progress and integrated management of frosty pod rot (*Moniliophthora roreri*) of cocoa in Tabasco, Mexico. *Journal of Plant Pathology* 93: 31–36.

De Souza, JT, Bailey, BA, Pomella, AWV, Erbe, EF, Murphy, CA, Bae, H, and Hebbar, PK. 2008. Colonization of cacao seedlings by *Trichoderma stromaticum*, a mycoparasite of the witches' broom pathogen, and its influence on plant growth and resistance. *Biological Control* 46: 36–45.

Díaz-Valderrama, JR, and Aime, MC. 2016. The cacao pathogen *Moniliophthora roreri* (Marasmiaceae) possesses biallelic A and B mating loci but reproduces clonally. *Heredity* 116: 491–501.

Dibog, L, Babin, R, Mbang, JMA, Decazy, B, Nyassé, S, Cilas, C, and Eskes, AB. 2008. Effect of genotype of cocoa (*Theobroma cacao* L.) on attractiveness to the mirid *Sahlbergella singularis* (Hemiptera: Miridae) in the laboratory. *Pest Management Science* 64: 977–980.

Djocgoue, PF, Boudjeko, T, Mbouobda, HD, Nankeu, DJ, El Hadrami, I, and Omokolo, ND. 2007. Heritability of phenols in the resistance of *Theobroma cacao* against *Phytophthora megakarya*, the causal agent of black pod disease. *Journal of Phytopathology* 155: 519–525.

Domfeh, O, Ameyaw, GA, Dzahini-Obiatey, HK, Ollennu, LAA, Osei-Bonsu, K, Acheampong, K, Aneani, F, and Owusu-Ansah, F. 2016. Use of immune crops as barrier in the management of cacao swollen shoot virus disease (CSSVD)—long-term assessment. *Plant Disease* 100: 1889–1893.

do Rio, MCS, de Oliveira, BV, de Thomazella, DPT, da Silva, JAF, and Pereira, GAG. 2008. Production of calcium oxalate crystals by the basidiomycete *Moniliophthora perniciosa*, the causal agent of witches' broom disease of cacao. *Current Microbiology* 56: 363–370.

Doudna, JA, and Charpentier, E. 2014. The new frontier of genome engineering with CRISPR-Cas9. *Science* 346: 1077. Doi: 10.1126/science.1258096.

Dufour, PS. 1671. *De l'Usage du Caphé, du Thé et du Chocolate*. Lyon, France: Iean Girin & Barthelemy Riviere. Translated into English as Dufour, PS. 1685. *The manner of Making Coffee, Tea and Chocolate as it is Used in Most Parts of Europe, Asia, Africa and America, with their Vertues*. London: W. Crook. Cited in Wilson, PK, and Hurst, WJ. 2012. *Chocolate as Medicine: a quest over the centuries*. Cambridge, UK: The Royal Society of Chemistry, p 57.

Eitzinger, A, Farrell, A, Rhiney, K, Carmona, S, van Loosen, I, and Taylor, M. 2015. "Trinidad and Tobago: assessing the impact of climate change on cocoa and tomato." CIAT Policy Brief No. 27. Cali, Colombia: Centro Internacional de Agricultura Tropical (CIAT).

Engelbrecht, CJ, Harrington, TC, and Alfenas, AC. 2007. Ceratocystis wilt of cacao—a disease of increasing importance. *Phytopathology* 97: 1648–1649.

Erneholm, I. 1948. "Cacao production of South America." Gothenburg, Sweden. Cited by Wood, GAR, and Lass, RA. 1985. *Cocoa*. Fourth Edition. Harlow, UK: Longman, p 52.

Evans, HC. 1973a. New developments in black pod epidemiology. *Cocoa Growers Bulletin* 20: 10–16.

Evans, HC. 1973b. Invertebrate vectors of *Phytophthora palmivora*, causing black pod disease of cocoa in Ghana. *Annals of Applied Biology* 75: 331–345.

Evans, HC. 1978. Witches' broom disease of cocoa (*Crinipellis perniciosa*) in Ecuador. *Annals of Applied Biology* 89: 185–192.

Evans, HC. 1980. Pleomorphism in *Crinipellis perniciosa*, causal agent of witches' broom disease of cocoa. *Transactions of the British Mycological Society* 74: 515–526.

Evans, HC. 1981. Witches' broom disease—a case study. *Cocoa Growers Bulletin* 32: 5–19.

Evans, HC. 2012. Cacao diseases in the Americas: myths and misnomers. *Fungi* 5 (4): 29–35.

Evans, HC. 2016. "Witches' broom disease (*Moniliophthora perniciosa*): history and biology." In *Cacao diseases: a history of old enemies and new encounters*, edited by BA Bailey and LW Meinhardt, 137–177. Switzerland: Springer International.

Evans, HC, Krauss, U, Rios, RR, Zecevich, TA, and Arévalo-Gardini, E. 1998. Cocoa in Peru. *Cocoa Growers' Bulletin* 51: 7–22.

Evans, HC, Holmes, KA, Thomas, SE (2003). Endophytes and mycoparasites associated with an indigenous forest tree, *Theobroma gileri*, in Ecuador and a preliminary assessment of their potential as biocontrol agents of cocoa diseases. *Mycological Progress* 2: 149–160.

Evans, HC, Bezerra, JL, and Barreto, RW. 2013. Of mushrooms and chocolate trees: aetiology and phylogeny of witches' broom and frosty pod diseases of cacao. *Plant Pathology* 62: 728–740.

FAO (Food and Agriculture Organisation). n.d. "Chocolate Stats." Accessed November 28, 2018. www.fao.org/assets/infographics/FAO-Infographic-chocolate-en.pdf.

FAO (Food and Agriculture Organisation). 2007. "Crop ecological requirements database (ECOCROP). Rome, Italy: Food and Agriculture Organization of the United Nations." http://www.fao.org/land-water/land/land-governance/land-resources-planning-toolbox/category/details/en/c/1027491/

Farand, C. 2018. "Carbon dioxide levels in Earth's atmosphere reach highest level in 800,000 years." *Independent*, May 5, 2018. https://www.independent.co.uk/environment/carbon-dioxide-concentration-atmosphere-highest-level-800000-years-mauna-loa-observatory-hawaii-a8337921.html.

Farrell, AD, Rhiney, K, Eitzinger, A, and Umaharan, P. 2018. Climate adaptation in a minor crop species: is the cocoa breeding network prepared for climate change? *Agroecology and Sustainable Food Systems* 42(7): 812–833.

Felicitas, AC, Hervé, BDB, Ekesi, S, Akutse, KS, Djuideu, CTCL, Meupia, MJ, and Babalola, OO. 2018. Consequences of shade management on the taxonomic patterns and functional diversity of termites (Blattodea: Termitidae) in cocoa agroforestry systems. *Ecology and Evolution* 8: 11582–11595.

Fernandes, LDS, Royaert, S, Corrêa, FM, Mustiga, GM, Marelli, J-P, Corrêa, RX, and Motamayor, JC. 2018. Mapping of a major QTL for *Ceratocystis* wilt disease in an F1 population of *Theobroma cacao*. *Frontiers in Plant Science* 9: 155. Doi: 10.3389/fpls.2018.00155

Fiorin, GL, Sanchéz-Vallet, A, Thomazella, DPT, do Prado, PFV, do Nascimento, LC, Figueira, AVO, Thomma, BPHJ, Pereira, GAG, and Teixeira, PJPL. 2018. Suppression of plant immunity by fungal chitinase-like effectors. *Current Biology* 28: 1–8.

Fister, AS, Landherr, L, Maximova, SN, and Guiltinan, MJ. 2018. Transient expression of CRISPR/Cas9 machinery targeting TcNPR3 enhances defense response in *Theobroma cacao*. *Frontiers in Plant Science* 9: article 268. Doi: 10.3389/fpls.2018.00268

Flood, J, and Gilmour, M. 2017. "The potential effects of climate change on cacao pests and diseases." International Indonesian Cocoa Symposium, October 18–20, 2017, Nusa-Dua, Bali, Indonesia.

Forbes, SJ. 2015. *Increased pollinator habitat enhances cacao fruit set and predator conservation in Northern Australia*. MSc Thesis, James Cook University, Cairns, Queensland, Australia.

Forbes, SJ, and Northfield, TD. 2017a. *Oecophylla smaragdina* ants provide pest control in Australian cacao. *Biotropica* 49: 328–336.

Forbes, SJ, and Northfield, TD. 2017b. Increased pollinator habitat enhances cacao fruit set and predator conservation. *Ecological Applications* 27: 887–899.

Ford, CS, and Wilkinson, MJ. 2012. Confocal observations of late-acting self-incompatibility in *Theobroma cacao* L. *Sexual Plant Reproduction* 25: 169–183.

Ford, T, Vit, J, Neate, R, Branigan, T, and Saner, E. 2014. "The cocoa crisis: why the world's stash of chocolate is melting away." *The Guardian*, November 21, 2014. https://www.theguardian.com/lifeandstyle/2014/nov/21/cocoa-crisis-world-chocolate-stash-melting-away.

Francis, ST, Head, K, Morris, PG, and Macdonald, IA. 2006. The effect of flavanol-rich cocoa on the fMRI response to a cognitive task in healthy young people. *Journal of Cardiovascular Pharmacology* 47: S215–S220.

Frauendorfer, F, and Schieberle, P. 2006. Identification of the key aroma compounds in cocoa powder based on molecular sensory correlations. *Journal of Agricultural and Food Chemistry* 54: 5521–5529.

Frias, GA, and Purdy, LH. 1991. Infection biology of *Crinipellis perniciosa* on vegetative flushes of cacao. *Plant Disease* 75: 552–556.

Fuller, LK. 1994. *Chocolate fads, folklore & fantasies: 1,000+ chunks of chocolate information*. New York: Haworth Press.

Gadsby, P. 2002. Endangered chocolate. *Discover* 23(8): 64–71.

Gardella, DS, Enriquez, GA, and Saunders, JL. 1982. Inheritance of clonal resistance to *Ceratocystis fimbriata* in cacao hybrids. *Proceedings of the 8th International Cocoa Research Conference*, Cartagena, Colombia, October 18–23, 1981. Colombia Cocoa Producers' Alliance, 695–702.

Gateau-Rey, L, Tanner, EVJ, Rapidel, B, Marelli, J-P, and Royaert, S. 2018. Climate change could threaten cocoa production: effects of 2015–16 El Niño-related drought on cocoa agroforests in Bahia, Brazil. *PLOS ONE* 13(7): e0200454. https://doi.org/10.1371/journal.pone.0200454.

Gidoin, C, Babin, R, Beilhe, LB, Cilas, C, ten Hoopen, GM, Ngo, Bieng, MA. 2014. Tree spatial structure, host composition and resource availability influence mirid density or black pod prevalence in cacao agroforests in Cameroon. *PLOS ONE* 9(10): e109405. Doi:10.1371/journal.pone.0109405

Gockowski, J, and Sonwa, D. 2011. Cacao intensification scenarios and their predicted impact on CO_2 emissions, biodiversity conservation, and rural livelihoods in the Guinea rainforest of West Africa. *Environmental Management* 48: 307–321.

Gorenz, AM. 1969. Spread of *Phytophthora* pod rot from the tree base to pods in the canopy. *Annual Report of the Cocoa Research Institute of Nigeria 1968–69*: 53–54.

Gras, P, Tscharntke, T, Maas, B, Tjoa, A, Hafsah, A, and Clough, Y. 2016. How ants, birds and bats affect crop yield along shade gradients in tropical cacao agroforestry. *Journal of Applied Ecology* 53: 953–963.

Green, M. 2017. How the decadence and depravity of London's 18th century elite was fuelled by hot chocolate. *Telegraph*, March 11, 2017.

Griffith, GW. 2007. *Phytophthora*: a blight on Ireland. *Microbiology Today*, February 7: 13–15.

Griffith, GW, and Hedger, JN. 1994. Spatial distribution of mycelia of the liana (L-) biotype of the agaric *Crinipellis perniciosa* (Stahel) Singer in tropical forest. *New Phytologist* 127: 243–259.

Guest, D. 2007. Black pod: diverse pathogens with a global impact on cocoa yield. *Phytopathology* 97: 1650–1653.

Guiltinan-Maximova Lab. n.d. Lab website. Accessed July 31, 2019. https://plantscience.psu.edu/research/labs/guiltinan.

Guiraud, BSHB, Tahi, MG, Fouet, O, Trebissou, CI, Pokou, D, Rivallan, R, Argout, X, Koffi, KK, Koné, B, Zoro, BIA, and Lanaud, C. 2018. Assessment of genetic diversity and structure in cocoa trees (*Theobroma cacao* L.) in Côte d'Ivoire with reference to their susceptibility to cocoa swollen shoot virus disease (CSSVD). *Tree Genetics and Genomes* 14: 52. https://doi.org/10.1007/s11295-018-1264-y

Gullan, PJ, and Cranston, PS. 2014. *The Insects. An outline of entomology.* Oxford, UK: Wiley-Blackwell.

Gutiérrez, TJ. 2017. State-of-the-art chocolate manufacture: a review. *Comprehensive Reviews in Food Science and Food Safety* 16: 1313–1344.

Head, B. 1903. *The food of the gods: a popular account of cocoa.* London: R. Brimley Johnson.

Heffer Link, V, Powelson, ML, and Johnson, KB. 2002. Oomycetes. *Plant Health Instructor*. DOI: 10.1094/PHI-I-2002-0225-01. https://www.apsnet.org/edcenter/intropp/LabExercises/Pages/Oomycetes.aspx

Henson, R. 2008. *The rough guide to climate change: the symptoms, the science, the solutions.* London: Rough Guides.

Ho, VTT, Zhao, J, and Fleet, G. 2015. The effect of lactic acid bacteria on cocoa bean fermentation. *International Journal of Food Microbiology* 205: 54–67.

Holliday, P. 1953. The cultivated cacao of Colombia. *Journal of the Agricultural Society of Trinidad and Tobago* 53: 397–406.

Holmes, KA, Evans, HC, Wayne, S, and Smith, J. 2003. *Irvingia*, a forest host of the cocoa black-pod pathogen, *Phytophthora megakarya*, in Cameroon. *Plant Pathology* 52: 486–490.

Hosonuma, N, Herold, M, De Sy, V, de Fries, RS, Brockhaus, M, Verchot, L, Angelsen, A, and Romijn, E. 2012. An assessment of deforestation and forest degradation drivers in developing countries. *Environmental Research Letters* 7(4): 44009. https://doi.org/10.1088/1748-9326/7/4/044009.

Hughes, W. 1672. *The American Physitian or a Treatise of the Roots, Plants, Trees, Shrubs, Fruit, Herbs &c. Growing in the English Plantations in America: Describing the Place, Time, Names, Kindes, Temperature, Vertues and Uses of them, either for Diet, Physick, &c. Whereunto is added A Discourse of the Cacao-nut Tree, and the use of its Fruit; with all the ways of making Chocolate. The like never extant before.* London: J.C. for William Crook. Cited in Wilson, PK, and Hurst, WJ. 2012. *Chocolate as medicine: a quest over the centuries.* Cambridge, UK: Royal Society of Chemistry, p 53.

Hurst, WJ, Tarka, SM, Powis, TG, Valdez, F, and Hester, TR. 2002. Archaeology: Cacao usage by the earliest Maya civilization. *Nature* 418: 289–290.

[ICCO] International Cocoa Organisation. 2010. "Impact of El Niño and La Niña weather events." https://www.icco.org/about-us/international-cocoa-agreements/cat_view/30-related-documents/31-world-cocoa-market.html

[ICCO] International Cocoa Organisation. 2012. "How many smallholders are there worldwide producing cocoa?" Accessed November 1, 2019. https://www.icco.org/faq/57-cocoa-production/123-how-many-smallholders-are-there-worldwide-pro-ducing-cocoa-what-proportion-of-cocoa-worldwide-is-produced-by-smallholders.html.

[ICCO] International Cocoa Organisation. 2014. "Zurich Certification Workshop finds common ground." Accessed July 12, 2019. http://www.icco.org/about-us/icco-news/253-zurich-certification-workshop-finds-common-ground.html.

[ICCO] International Cocoa Organization. 2015. "Pest and diseases. Cocoa pod borer (*C. cramerella*)." Accessed July 19 2019. https://www.icco.org/about-cocoa/pest-a-diseases.

[ICCO] International Cocoa Organisation. 2018. *Quarterly Bulletin of Cocoa Statistics* 44(3), Cocoa Year 2017–18.

[ICCO] International Cocoa Organization. 2019. "The chocolate Industry." Accessed June 25, 2019. https://www.icco.org/about-cocoa/chocolate-industry.html.

Innovative Genomics Institute. n.d. 2018. "New Project to Create Disease-Resistant Ca-cao." Accessed July 31, 2019. https://innovativegenomics.org/news/new-project-to-create-disease-resistant-cacao/.

International Cocoa Initiative. 2017. "Putting children first." Annual Report 2017. https://annualreport2017.cocoainitiative.org/

[IPCC] Intergovernmental Panel on Climate Change. 2014. Climate change 2014. "Synthesis Report, Summary for Policymakers." https://www.ipcc.ch/report/ar5/syr/

Isman, MB. 2006. Botanical insecticides, deterrents, and repellents in modern agriculture and an increasingly regulated world. *Annual Review of Entomology* 51: 45–66.

Iwaro, AD, Sreenivasan, TN, and Umaharan, P. 1997. *Phytophthora* resistance in cacao (*Theobroma cacao*): influence of pod morphological characteristics. *Plant Pathology* 46: 557–565.

Jagoret, P, Michel-Dounias, I, and Malézieux, E. 2011. Long-term dynamics of cocoa agroforests: a case study in central Cameroon. *Agroforestry Systems* 81(3): 267–278. https://doi.org/10.1007/s10457-010-9368-x.

Jagoret, P, Michel-Dounias, I, Snoeck, D, Ngnogué, HT, and Malézieux, E. 2012. Afforestation of savannah with cocoa agroforestry systems: a small farmer innovation in central Cameroon. *Agroforestry Systems* 86: 493–504. https://doi.org/10.1007/s10457-012-9513-9.

Jagoret, P, Ngnogue, HT, Malézieux, E, and Michel, I. 2018. Trajectories of cocoa agroforests and their drivers over time: lessons from the Cameroonian experience. *European Journal of Agronomy* 101: 183–192.

Johns, ND. 1999. Conservation in Brazil's chocolate forest: the unlikely persistence of the traditional cocoa agroecosystem. *Environmental Management* 23: 31–47.

Johnson, SN, and Gregory, PJ. 2006. Chemically-mediated host-plant location and selection by root-feeding insects. *Physiological Entomology* 31: 1–13.

Jorgensen, H. 1970. *Monilia* pod rot of cacao in Ecuador. *Turrialba* 15: 4–13.

Jouquet, P, Dauber, J, Lagerlof, J, Lavelle, P, and Lepage, M. 2006. Soil invertebrates as ecosystem engineers: intended and accidental effects on soil and feedback loops. *Applied Soil Ecology* 32: 153–164.

Jouquet, P, Traoré, S, Choosai, C, Hartmann, C, and Bignell, D. 2011. Influence of termites on ecosystem functioning. Ecosystem services provided by termites. *European Journal of Soil Biology* 47: 215–222.

Karp, DS, Mendenhall, CD, Sandí, RF, Chaumont, N, Ehrlich, PR, Hadly, EA, and Daily, GC. 2013. Forest bolsters bird abundance, pest control and coffee yield. *Ecology Letters* 16: 1339–1347.

Keane, PJ. 1981. Epidemiology of vascular-streak dieback of cocoa. *Annals of Applied Biology* 98: 227–241.

Keane, PJ. 2010. Lessons from the tropics—plant diversity, unusual and changeable plant pathology, horizontal resistance, and the plight of farmers. *Australasian Plant Pathology* 39: 192–201.

Keane, PJ, Flentje, MT, and Lamb, KP. 1972. Investigation of vascular-streak dieback of cocoa in Papua New Guinea. *Australian Journal of Biological Sciences* 25: 553–564.

Keane, PJ, and Prior, C. 1991. Vascular-streak dieback of cocoa. *Phytopathological Papers* 33. Mycological Institute IV Series. Egham, Surrey, UK: CAB International.

Kew Science. n.d. "Plants of the world online: *Theobroma cacao* L." http://powo.science. kew.org/taxon/urn:lsid:ipni.org:names:320783-2

Kühn, J, Schröter, A, Hartmann, BM, and Stangl, GI. 2018. Cocoa and chocolate are sources of vitamin D2. *Food Chemistry* 269: 318–320.

Lachenaud, P, and Mossu, G. 1985. Etude comparative de l'influence de deux modes de conduite sur les facteurs du rendement d'une cacaoyere. *Café Cacao Thé* 29(1): 21–30.

Lachenaud, P, Rossi, V, Thevenin, J-M, and Doaré, F. 2015. The 'Guiana' genetic group: a new source of resistance to cacao (*Theobroma cacao* L.) black pod rot caused by *Phytophthora capsici*. *Crop Protection* 67: 91–95.

Läderach, P, Martinez-Valle, A, Schroth, G, and Castro, N. 2013. Predicting the future climatic suitability for cocoa farming of the world's leading producer countries, Ghana and Côte d'Ivoire. *Climatic Change* 119: 841–854.

Lahive, F, Hadley, P, Daymond, AJ. 2019. The physiological responses of cacao to the environment and the implications for climate change resilience. A review. *Agronomy for Sustainable Development* 39: article 5.

Lamuela-Raventós, RM, Romero-Pérez, AI, Andrés-Lacueva, C, and Tornero, A. 2005. Review: health effects of cocoa flavonoids. *Food Science and Technology International* 11(3): 159–176.

Lee, BY. 2018. "Could chocolate actually go extinct in 40 years?" *Forbes*, January 1, 2018. https://www.forbes.com/sites/brucelee/2018/01/01/is-chocolate-going-extinct-whats-being-done-to-prevent-this-disaster/#5ad4569d540f.

Legg, JT. 1982. *The Cocoa Swollen Shoot Research Project at the Cocoa Research Institute, Tafo, Ghana 1969–78*. London: Overseas Development Administration.

Leissle, K. 2018. *Cocoa*. Cambridge, UK: Polity Press.

Lémery, N. 1745. *A Treatise of all Sorts of Foods, both Animal and Vegetable: Also of Drinkables: Giving an Account How to Chuse the Best Sort of all Kinds*. Translated by Hay, D, Innys, W, Longman, T, and Shewell, T. London. Cited in Wilson, PK, and Hurst, WJ. 2012. *Chocolate as medicine: a quest over the centuries*. Cambridge, UK: Royal Society of Chemistry, p. 61.

León-Portilla, M. 2002. *Bernadino de Sahagún: the first anthropologist*. Norman, OK, USA: University of Oklahoma Press.

LSE (The London School of Economics and Political Science / Grantham Research Institute on Climate Change and the Environment). 2018. "What is the role of deforestation in climate change and how can Reducing Emissions from Deforestation and Degradation (REDD+) help?" Accessed November 28, 2018. http://www.lse.ac.uk/GranthamInstitute/faqs/whats-redd-and-will-it-help-tackle-climate-change/.

MacLean, JAR. 1953. Some chemical aspects of 'black pod' disease in West African Amelonado cacao. *Empire Journal of Experimental Agriculture* 21: 340–349.

Maclean, R. 2017. "Chocolate industry drives rainforest disaster in Ivory Coast." *Guardian*, September 13, 2017. https://www.theguardian.com/environment/2017/sep/13/chocolate-industry-drives-rainforest-disaster-in-ivory-coast.

Maclean, R. 2018. "Africa cocoa industry failing on deforestation pledge–campaigners." *Guardian*, December 7, 2018. https://www.theguardian.com/environment/2018/dec/07/africa-cocoa-industry-failing-deforestation-pledge-campaigners?CMP=Share_iO-SApp_Other.

Maddison, AC, and Griffin, MJ. 1981. "Detection and movement of inoculum." In *Epidemiology of* Phytophthora *on cocoa in Nigeria*, edited by PH Gregory and AC Maddison, 31–50. Phytopathological Paper 25. Kew, London: Commonwealth Mycological Institute.

Maddison, AC, and Ward, MR. 1981. The international black pod project: 1979 review. *Proceedings of the 7th International Cocoa Research Conference*, Douala, Cameroon, November 5–12, 261–266.

Madge, DS. 1968. The behaviour of the cocoa mirid (*Sahlbergella singularis* Hagl.) to some environmental factors. *Bulletin of the Entomological Society of Nigeria* 1: 63–70.

Magagna, F, Guglielmetti, A, Liberto, E, Reichenbach, SE, Allegrucci, E, Gobino, G, Bicchi, C, and Cordero, C. 2017. Comprehensive chemical fingerprinting of high-quality cocoa at early stages of processing: effectiveness of combined untargeted and targeted approaches for classification and discrimination. *Journal of Agricultural and Food Chemistry* 65: 6329–6341.

Mahob, RJ, Feudjio Thiomela, R, Dibog, L, Babin, R, Fotso Toguem, YG, Mahot, H, Baleba, L, Owona Dongo, PA, and Bilong Bilong, CF. 2019. Field evaluation of the impact of *Sahlbergella singularis* Haglund infestations on the productivity of different *Theobroma cacao* L. genotypes in the Southern Cameroon. *Journal of Plant Diseases and Protection* 126: 203–210.

Make Chocolate Fair. n.d. "Cocoa production in a nutshell." Accessed October 31, 2019. https://makechocolatefair.org/issues/cocoa-production-nutshell.

Marsh, CE, Carter, HH, Guelfi, KJ, Smith, KJ, Pike, KE, Naylor, LH, and Green, DJ. 2017. Brachial and cerebrovascular functions are enhanced in postmenopausal women after ingestion of chocolate with a high concentration of cocoa. *Journal of Nutrition* 147: 1686–1692.

Martín, MA, and Ramos, S. 2017. Health beneficial effects of cocoa phenolic compounds: a mini-review. *Current Opinion in Food Science* 14: 20–25.

Maas, B, Clough, Y, and Tscharntke, T. 2013. Bats and birds increase crop yield in tropical agroforestry landscapes. *Ecological Letters* 16: 1480–1487.

Mbenoun, M, Momo Zeutsa, EH, Samuels, G, Nsouga Amougou, F, and Nyasse, S. 2008. Dieback due to *Lasiodiplodia theobromae*, a new constraint to cocoa production in Cameroon. *Plant Pathology* 57(2): 381.

McMahon, P, and Purwantara, A. 2016. "Vascular streak dieback (*Ceratobasidium theobromae*): history and biology." In *Cacao diseases: a history of old enemies and new encounters*, edited by BA Bailey and LW Meinhardt, 307–335. Switzerland: Springer International.

McMahon, PJ, Susilo, AW, Parawansa, AK, Bryceson, SR, Nurlaila, MS, Saftar, A, Purwantara, A, Purung, HB, Lambert, S, Guest, DI, and Keane, PJ. 2018. Testing local cacao selections in Sulawesi for resistance to vascular streak dieback. *Crop Protection* 109: 24–32.

McVitie's. n.d. "About McVitie's." Accessed April 20, 2018. http://www.mcvities.co.uk/about.

Meinhardt, LW, Rincones, J, Bailey, BA, Aime, MC, Griffith, GW, Zhang, D, and Pereira, GAG. 2008. *Moniliophthora perniciosa*, the causal agent of witches' broom disease of cacao: what's new from this old foe? *Molecular Plant Pathology* 9: 577–588.

Melis, M, Carboni, E, Caboni, P, and Acquas, E. 2015. Key role of salsolinol in ethanol actions on dopamine neuronal activity of the posterior ventral tegmental area. *Addiction Biology* 20: 182–193.

Melzig, MF, Putscher, I, Henklein, P, and Haber, H. 2000. *In vitro* pharmacological activity of the tetrahydroisoquinoline salsolinol present in products from *Theobroma cacao* L. like cocoa and chocolate. *Journal of Ethnopharmacology* 73: 153–159.

Met Office. n.d. "What is climate change?" Accessed September 27, 2018. https://www.metoffice.gov.uk/climate-guide/climate-change.

Mexicolore. n.d. "How to cure the stiffening of the serpent: herbal medicine, Aztec style." Accessed April 20, 2018. http://www.mexicolore.co.uk/aztecs/health/herbal-medicine-aztec-style.

Micheli, F, Guiltinan, MJ, Gramacho, KP, Wilkinson, MJ, Figueira, AVO, Cascardo, JCM, Maximova, SN, and Lanaud, C. 2010. Functional genomics of cacao. *Advances in Botanical Research* 55: 119–177.

Mighty Earth. n.d. "Your cocoa, kissed by deforestation." Accessed December 7, 2018. http://www.mightyearth.org/kissed-by-deforestation/.

Mitchard, ETA. 2018. The tropical forest carbon cycle and climate change. *Nature* 559: 527–534.

Miyaji, K-I, da Silva, WS, and Alvim, P de T. 1997a. Productivity of leaves of a tropical tree, *Theobroma cacao*, grown under shading, in relation to leaf age and light conditions within the canopy. *New Phytologist* 137(3): 463–472. https://doi.org/10.1046/j.1469-8137.1997.00841.x

Miyaji, K-I, da Silva, WS, Alvim, P de T. 1997b. Longevity of leaves of a tropical tree, *Theobroma cacao*, grown under shading, in relation to position within the canopy and time of emergence. *New Phytologist* 137(3): 445–454. https://doi.org/10.1046/j.1469-8137.1997.00667.x

Moguel, P, and Toledo, VM. 1999. Biodiversity conservation in traditional coffee systems of Mexico. *Conservation Biology* 23: 31–47.

Mondego, JMC, Thomazella, DPT, Teixeira, PJPL, and Pereira, GAG. 2016. "Genomics, transcriptomics, and beyond: the fifteen years of cacao's witches' broom disease genome project." In *Cacao diseases: a history of old enemies and new encounters*, edited by BA Bailey and LW Meinhardt, 179–210. Switzerland: Springer International.

Money, NP. 2007. *The triumph of the fungi*. Oxford, UK: Oxford University Press.

Mortimer, R, Saj, S, and David, C. 2018. Supporting and regulating ecosystem services in cacao agroforestry systems. *Agroforestry Systems* 92: 1639–1657. https://doi.org/10.1007/s10457-017-0113-6.

Morton, M, and Morton, F. 1986. *Chocolate: an illustrated history*. New York: Crown Publishers.

Motamayor, JC, Lachenaud, P, da Silva e Mota, J, Loor, R, Kuhn, DN, Brown, JS, and Schnell, RJ. 2008. Geographical and genetic population differentiation of the Amazonian chocolate tree (*Theobroma cacao* L). *PLOS ONE* 3(10): e3311. doi:10.1371/journal.pone.0003311.

Motilal, L, and Butler, D. 2003. Verification of identities in global cacao germplasm collections. *Genetic Resources and Crop Evolution* 50: 799–807.

Muller, E. 2016. "Cacao swollen shoot virus (CSSV): history, biology, and genome." In *Cacao diseases: a history of old enemies and new encounters*, edited by BA Bailey and LW Meinhardt, 337–358. Switzerland: Springer International.

Muller, E, Ravel, S, Agret, C, Abrokwah, F, Dzahini-Obiatey, H, Galyuon, I, Kouakou, K, Jeyaseelan, EC, Allainguillaume, J, and Wetten, A. 2018. Next generation sequencing elucidates cacao badnavirus diversity and reveals the existence of more than ten viral species. *Virus Research* 244: 235–251.

Naipaul, Sir Vidia. 2011. *A way in the world: a sequence*. London: Picador.

NASA. n.d. "Global Climate Change. Vital signs of the planet." Accessed September 27, 2018. https://climate.nasa.gov/evidence/.

NASA. n.d. "Science, Solar System Exploration: Mars." Accessed July 12, 2019. https://solarsystem.nasa.gov/planets/mars/in-depth/

Ndoye, O, Perez, MR, and Eyebe, A. 1998. "Non-timber forest products in the humid forest zone of Cameroon." Rural Development Forestry Network paper 22c. London: Overseas Development Institute.

Nicol, J. 1953. The capsid problem. *Proceedings of the West African International Cacao Research Conference*, December 12–16, 1953, West African Cacao Research Institute,

Tafo, Gold Coast. London: Crown Agents for Overseas Governments and Administrations for the Conference Committee. 51–52.

Nieburg, O. 2014. "Paying the price of chocolate: breaking cocoa farming's cycle of poverty." *Confectionery News*. Accessed November 28, 2018. https://www.confectionerynews.com/Article/2014/07/10/Price-of-Chocolate-Breaking-poverty-cycle-in-cocoafarming.

Nieburg, O. 2018. "How will the chocolate industry approach cocoa farmer 'living income'?" *Confectionery News*. Accessed September 27, 2018. https://www.confectionerynews.com/Article/2018/05/03/How-will-the-chocolate-industry-approach-cocoafarmer-living-income.

Niether, W, Armengot, L, Andres, C, Schneider, M, and Gerold, G. 2018. Shade trees and tree pruning alter throughfall and microclimate in cocoa (*Theobroma cacao* L.) production systems. *Annals of Forest Science* 75: 38. https://doi.org/10.1007/s13595-018-0723-9

Nunoo, I, and Owusu, V. 2017. Comparative analysis on financial viability of cocoa agroforestry systems in Ghana. *Environment, Development and Sustainability* 19: 83–98.

Nyadanu, D, Akromah, R, Adomako, B, Kwoseh, C, Dzahini-Obiatey, H, Lowor, ST, Akrofi, AY, and Assuah, MK. 2012. Host plant resistance to *Phytophthora* pod rot in cacao (*Theobroma cacao* L.): the role of epicuticular wax on pod and leaf surfaces. *International Journal of Botany* 8: 13–21.

Nyadanu, D, Lowor, ST, Akrofi, AY, Adomako, B, Dzahini-Obiatey, H, Akromah, R, Awuah, RT, Kwoseh, C, Adu-Amoah, R, and Kwarteng, AO. 2019. Mode of inheritance and combining ability studies on epicuticular wax production in resistance to black pod disease in cacao (*Theobroma cacao* L.). *Scientia Horticulturae* 243: 34–40.

Nyeko, P, and Olubayo, FM. 2005. Participatory assessment of farmers' experiences of termite problems in agroforestry in Tororo District, Uganda. *Agricultural Research and Extension Network (AgREN)*. London: Overseas Development Institute (ODI).

Off, C. 2006. *Bitter chocolate: anatomy of an industry*. New York: New Press.

Offenberg, J. 2015. Ants as tools in sustainable agriculture. *Journal of Applied Ecology* 52: 1197–1205.

Ofori, A, Padi, FK, Ameyaw, GA, Dadzie, AM, and Lowor, S. 2015. Genetic variation among cocoa (*Theobroma cacao* L.) families for resistance to cocoa swollen shoot virus disease in relation to total phenolic content. *Plant Breeding* 134: 477–484.

Ogedegbe, AB, and Ogwu, BC. 2015. Termite infestation on farmlands at Ugoniyekorhionmwon Community, Edo State, Nigeria: a farmers' perception. *International Journal of Pure and Applied Sciences and Technology* 28: 8.

Okaisabor, EK. 1971. The mechanism of initiation of *Phytophthora* pod rot epiphytotics. *Proceedings of the Third International Cocoa Research Conference*, Accra, Ghana, 1969: 398–404.

Opoku, IY, Akrofi, AY, and Appiah, AA. 2002. Shade trees are alternative hosts of the cocoa pathogen *Phytophthora megakarya*. *Crop Protection* 21: 629–634.

Ozturk, G, and Young, GM. 2017. Food evolution: the impact of society and science on the fermentation of cocoa beans. *Comprehensive Reviews in Food Science and Food Safety* 16: 431–456.

Padi, B. 1997. Prospects for the control of cocoa capsids—alternatives to chemical control. *Proceedings of the 1st International Cocoa Pests and Diseases Seminar*, November 6–10, 1995, Accra, Ghana. 28–36.

Padi, B, and Owusu, GK. 1998. Towards an integrated pest management for sustainable cocoa production in Ghana. *Proceedings of the 1st Sustainable Cocoa Workshop*, Panama, March 20–April 3, 1998, 7–15.

Perfecto, I, and Castiñeiras, A. 1998. "Deployment of the predaceous ants and their conservation in agroecosystems." In *Conservation biological control*, edited by P Barbosa, 269–289. London: Academic Press.

Perfecto, I, and Vandermeer, J. 2008. Biodiversity conservation in tropical agroecosystems: a new conservation paradigm. *Annals of the New York Academy of Sciences* 1134: 173–200.

Ploetz, R. 2016. "The impact of diseases on cacao production: a global overview." In *Cacao diseases: a history of old enemies and new encounters*, edited by BA Bailey and LW Meinhardt, 33–59. Switzerland: Springer International.

Posnette, AF. 1940. Transmission of swollen-shoot. *Tropical Agriculture (Trinidad)* 17: 98.

Posnette, Peter, obituary. 2004. *Telegraph* July 27, 2004. https://www.telegraph.co.uk/news/obituaries/1467914/Peter-Posnette.html

Pound, FJ. 1932. "The genetic constitution of the cacao crop." First Annual Report of Cacao Research 1931, Trinidad, 10–24.

Prior, C. 1980. Vascular-streak dieback. *Cocoa Growers Bulletin* 29: 21–26.

Prior, C. 1985. Cocoa quarantine: measures to prevent the spread of vascular-streak dieback in planting material. *Plant Pathology* 34: 603–608.

Purseglove, JW. 1968. *Tropical Crops: Dicotyledons*. London: Longman.

Raja Harun, RM, and Hardwick, K. 1988. The effects of prolonged exposure to different light intensities on the photosynthesis of cocoa leaves. *Proceedings of the 10th International Cocoa Research Conference*, Santo Domingo, Dominican Republic, May 17–23, 1987, 205–209.

Resende, MLV, Nojosa, GBA, Cavalcanti, LS, Aguilar, MAG, Silva, LHCP, Perez, JO, Andrade, GCG, Carvalho, GA, and Castro, RM. 2002. Induction of resistance in cocoa against *Crinipellis perniciosa* and *Verticillium dahliae* by acibenzolar-S-methyl (ASM). *Plant Pathology* 51: 621–628.

Resourcetrade.earth. n.d. "Cocoa trade, climate change and deforestation." Accessed September 27, 2018. https://resourcetrade.earth/stories.

Rice, RA, and Greenberg, R. 2000. Cacao cultivation and the conservation of biological diversity. *Ambio* 29: 167–173.

Rizali, A, Tscharntke, T, Buchori, D, and Clough, Y. 2018. Separating effects of species identity and species richness on predation, pathogen dissemination and resistance to invasive species in tropical ant communities. *Agricultural and Forest Entomology* 20: 122–130.

Rorer, JB. 1913. The Suriname witch-broom disease of cacao. *Circular of the Department of Agriculture of Trinidad and Tobago* 10: 1–13.

Rorer, JB. 1926. Ecuador cacao. *Tropical Agriculture (Trinidad)* 3(46–47): 68–69.

Rosmana, A, Samuels, GJ, Ismaiel, A, Ibrahim, ES, Haveri, P, Herawati, Y, and Asman, A. 2015. *Trichoderma asperellum*: a dominant endophyte species in cacao grown in Sulawesi with potential for controlling vascular streak dieback disease. *Tropical Plant Pathology* 40: 19–25.

Royaert, S, Jansen, J, da Silva, DV, Branco, SMJ, Livingstone, DS, Mustiga, G, Marelli, J-P, Araújo, IS, Corrêa, RX, and Motamayor, JC. 2016. Identification of candidate genes involved in witches' broom disease resistance in a segregating mapping population of *Theobroma cacao* L. in Brazil. *BMC Genomics* 17: 107. DOI 10.1186/s12864–016–2415-x.

Royal Society / National Academy of Sciences. 2014. "Climate change: evidence and causes." https://royalsociety.org/topics-policy/projects/climate-change-evidence-causes/

Ruf, FO. 2011. The myth of complex cacao agroforests: the case of Ghana. *Human Ecology* 39: 373–388.

Ruf, F, Schroth, G, and Doffangui, K. 2015. Climate change, cocoa migrations and deforestation in West Africa—what does the past tell us about the future? *Sustainability Science* 10: 101–111. http://dx.doi.org/10.1007/s11625–014–0282-4.

Ruppel, CD, and Kessler, JD. 2017. The interaction of climate change and methane hydrates. *Reviews of Geophysics* 55: 126–168.

Ryan, O. 2011. *Chocolate Nations: living and dying for cocoa in West Africa*. London: Zed Books.

Saj, S, Durot, C, Mvondo Sakouma, K, Tayo, K, and Avana-Tientcheu, ML. 2017. Contribution of associated trees to long-term species conservation, carbon storage and sustainability: a functional analysis of tree communities in cacao plantations of central Cameroon. *International Journal of Agricultural Sustainability* 15: 282–302.

Salazar, JCS, Ngo Bieng, MA, Melgarejo, LM, Di Rienzo, JA, and Casanoves, F. 2018. First typology of cacao (*Theobroma cacao* L.) systems in Colombian Amazonia, based on tree species richness, canopy structure and light availability. *PLOS One* 13(2): e0191003. https://doi.org/10.1371/journal.pone.0191003.

Sambuichi, RHR, Vidal, DB, Piasentin, FB, Jardim, JG, Viana, TG, Menezes, AA, Mello, DLN, Ahnert, D, and Baligar, VC. 2012. Cabruca agroforests in southern Bahia, Brazil: tree component, management practices and tree species conservation. *Biodiversity and Conservation* 21: 1055–1077.

Schmitz, H, and Shapiro, H-Y. 2015. The race to save chocolate. *Scientific American*. June 1, 2015. https://www.scientificamerican.com/article/the-race-to-save-chocolate/.

Schroth, G, Krauss, U, Gasparotto, L, Duarte Aguilar, JA, and Vohland, K. 2000. Pests and diseases in agroforestry systems of the humid tropics. *Agroforestry Systems* 50(3): 199–241.

Schroth, G, Läderach, P, Matinez-Valle, AI, Bunn, C, Jassogne, L. 2016. Vulnerability to climate change of cocoa in West Africa: patterns, opportunities and limitations to adaptation. *Science of the Total Environment* 556: 231–241.

Schroth, G, Läderach, P, Martinez-Valle, AI, and Bunn, C. 2017. From site-level to regional adaptation planning for tropical commodities: cocoa in West Africa. *Mitigation and Adaptation Strategies for Global Change* 22: 903–927.

Schumann, GL, and D'Arcy, CJ. 2012. *Hungry Planet: stories of plant diseases*. St Paul, MN, USA: American Phytopathological Society.

Schumann, M, Patel, A, and Vidal, S. 2013. Evaluation of an 'attract and kill' strategy for western corn rootworm larvae. *Applied Soil Ecology* 64: 178–189.

Sethi, S. 2017. "The greatest threat to chocolate isn't what you think." *Forbes*, October 30, 2017. https://www.forbes.com/sites/simransethi/2017/10/30/the-greatest-threat-to-chocolate-isnt-what-you-think/amp/.

Smallman, S, and Brown, K. 2012. "Witches' broom: the mystery of chocolate and bioterrorism in Brazil." Accessed June 21, 2018. https://www.introtoglobalstudies.com/2012/03/witches-broom-the-mystery-of-chocolate-and-bioterrorism-in-brazil/.

Sounigo, O, Coulibaly, N, Brun, L, N'Goran, J, Cilas, C, and Eskes, AB. 2003. Evaluation of resistance of *Theobroma cacao* L. to mirids in Côte d'Ivoire: results of comparative progeny trials. *Crop Protection* 22: 615–621.

Srinivasan, A. 2018. "What termites can teach us." *New Yorker*, September 17, 2018. https://www.newyorker.com/magazine/2018/09/17/what-termites-can-teach-us.

Stahel, G. 1915. *Marasmius perniciosus* nov. spec. *Bulletin Departement van den Landbouw in Suriname* 33: 5–26.

Stapley, JH, and Hammond, PS. 1957. Large-scale trials with insecticides against capsids on cocoa in Ghana. *Empirical Journal of Experimental Agriculture* 27: 343.

Statista.com. n.d. World cocoa production by country 2012/2013 to 2017/2018. Accessed November 11, 2019. https://www.statista.com/statistics/263855/cocoa-bean-production-worldwide-by-region/

Steffan-Dewenter, I, Kessler, M, Barkmann, J, Bos, M, Buchori, D, Erasmi, S, Faust, H, Gerold, G, Glenk, K, Gradstein, R, Guhardja, E, and others. 2007. Trade-offs between income, biodiversity, and ecosystem functioning during tropical rainforest conversion and agroforestry intensification. *Proceedings of the National Academy of Sciences USA* 104(12): 4973–4978.

Steven, WF. 1936. A new disease of cocoa in the Gold Coast. *Gold Coast Farmer* 5: 122, 144.

Stop the traffik. n.d. Accessed September 27, 2018. https://www.stopthetraffik.org/.

Stubbe, H. 1662. *The Indian Nectar, or a Discourse Concerning Chocolate wherein the Nature of the cacao-nut . . . is examined . . . the Ways of Compounding and Preparing chocolate are Enquired into; its effects, as to its Alimental and Venereal Quality, as well as Medicinal (Specially in Hypochondriacal Melancholy) are Fully Debated*. London: A. Crook. Cited in Wilson, PK, and Hurst, WJ. 2012. *Chocolate as Medicine: a quest over the centuries*. Cambridge, UK: Royal Society of Chemistry, p. 53.

Surujdeo-Maharaj, S, Sreenivasan, TN, Motilal, LA, and Umaharan, P. 2016. "Black pod and other *Phytophthora* induced diseases of cacao: history, biology, and control." In *Cacao diseases: a history of old enemies and new encounters*, edited by BA Bailey and LW Meinhardt, 213–266. Switzerland: Springer International.

Taylor, B, and Griffin, MJ. 1981. "The role and relative importance of different ant species in the dissemination of black pod disease of cocoa." In *Epidemiology of Phytophthora on cocoa in Nigeria*, edited by PH Gregory and AC Maddison, 114–131. Kew, London: Commonwealth Mycological Institute.

TCHO Chocolate. 2019. "Cocoa & the Living Income Differential: DMs with Laura Ann Sweitzer, Pt 2." Accessed October 24, 2019. https://umeshiso.com/post/582711042612/cocoa-the-living-income-differential-dms.

ten Hoopen, GM, and Krauss, U. 2016. "Biological control of cacao diseases." In *Cacao diseases: a history of old enemies and new encounters*, edited by BA Bailey and LW Meinhardt, 511–566. Switzerland: Springer International.

Teixeira, PJPL, Thomazella, DPT, and Pereira, GAG. 2015. Time for chocolate: current understanding and new perspectives on cacao witches' broom disease research. *PLOS Pathogens* 11(10): e1005130. doi:10.1371/journal.ppat.1005130.

Thevenin, J-M, Rossi, V, Ducamp, M, Doaré, F, Condina, V, and Lachenaud, P. 2012. Numerous clones resistant to *Phytophthora palmivora* in the 'Guiana' genetic group of *Theobroma cacao* L. *PLOS ONE* 7(7): e40915. doi:10.1371/journal.pone.0040915.

Thomazella, DPT, Teixeira, PJPL, Oliveira, HC, Saviani, EE, Rincones, J, Toni, IM, Reis, O, Garcia, O, Meinhardt, LW, Salgado, I, and Pereira, GAG. 2012. The hemibiotrophic cacao pathogen *Moniliophthora perniciosa* depends on a mitochondrial alternative oxidase for biotrophic development. *New Phytologist* 194: 1025–1034.

Thorlakson, T, and Neufeldt, H. 2012. Reducing subsistence farmers' vulnerability to climate change: evaluating the potential contributions of agroforestry in western Kenya. *Agriculture and Food Security* 1: 1–13.

Thresh, JM, Owusu, GK, and Ollennu, LA. 1988. Cocoa swollen shoot: an archetypal crowd disease. *Zeitschrift fur Pflanzenkrankheiten und Pflanzenschutz* 95: 428–446.

Tiburcio, RA, Costa, GG, Carazzolle, MF, Mondego, JM, Schuster, SC, Carlson, JE, Guiltinan, MJ, Bailey, BA, Mieczkowski, P, Meinhardt, LW, and Pereira, GA. 2010. Genes acquired by horizontal transfer are potentially involved in the evolution of phytopathogenicity in *Moniliophthora perniciosa* and *Moniliophthora roreri*, two of the major pathogens of cacao. *Journal of Molecular Evolution* 70: 85–97.

Toblerone. n.d. "Toblerone: how it all began—1900 The First Toblerone." Accessed November 12, 2018. http://www.toblerone.co.uk/history/howitbegan/1900.

Toledo-Hernández, M, Wanger, TC, and Tscharntke, T. 2017. Neglected pollinators: can enhanced pollination services improve cocoa yields? A review. *Agriculture, Ecosystems and Environment* 247: 137–148.

Tondoh, JE, Kouame, FNg, Martinez Guei, A, Sey, B, Wowo Kone, A, Gnessougou, N. 2015. Ecological changes induced by full-sun cocoa farming in Cote d'Ivoire. *Global Ecology and Conservation* 3: 575–595.

Tscharntke, T, Clough, Y, Bhagwat, SA, Buchori, D, Faust, H, Hertel, D, Holscher, D, Juhrbandt, J, Kessler, M, Perfecto, I, Scherber, C, Schroth, G, Veldkamp, E, and Wanger, TC. 2011. Multifunctional shade-tree management in tropical agroforestry landscapes—a review. *Journal of Applied Ecology* 48: 619–629.

Tuenter, E, Foubert, K, and Pieters, L. 2018. Mood components in cocoa and chocolate: the mood pyramid. *Planta Med* 84: 839–844.

Valenzuela, I, Purung, HB, Roush, RT, and Hamilton, AJ. 2014. Practical yield loss models for infestation of cocoa with cocoa pod borer moth, *Conopomorpha cramerella* (Snellen). *Crop Protection* 66: 19–28.

van Hall, CJJ. 1914. *Cocoa*. London: Macmillan.

van Hall, CJJ, and Drost, AW. 1909. The witch-broom disease in Suriname, its cause and treatment. *Proceedings of the Agricultural Society of Trinidad and Tobago* 9: 475–562.

Vanhove, W, Vanhoudt, N, and Van Damme, P. 2016. Biocontrol of vascular streak dieback (*Ceratobasidium theobromae*) on cacao (*Theobroma cacao*) through induced systemic resistance and direct antagonism. *Biocontrol Science and Technology* 26: 492–503.

van Wijngaarden, PM, van Kessel, M, and van Huis, A. 2007. *Oecophylla longinoda* (Hymenoptera: Formicidae) as a biological control agent for cocoa capsids (Hemiptera: Miridae). *Proceedings of the Netherlands Entomological Society* Meeting 18, 21–30. https://www.nev.nl/pages/publicaties/proceedings/proceeding_inhoud. php?nummer=18.

Vaast, P, and Somarriba, E. 2014. Trade-offs between crop intensification and ecosystem services: the role of agroforestry in cocoa cultivation. *Agroforestry Systems* 88: 947–956.

Veldkamp, E, Purbopuspito, J, Corre, MD, and Brumme, R. 2008. Land-use change effects on trace gases fluxes in the forest margins of Central Sulawesi, Indonesia. *Journal of Geophysical Research* 113: G02003. https://doi: 10.1029/2007JG000522.

Vlachojannis, J, Erne, P, Zimmermann, B, Chrubasik-Hausmann, S. 2016. The impact of cocoa flavanols on cardiovascular health. *Phytotherapy Research* 30: 1641–1657.

Vos, JGM, Ritchie, BJ, Flood, J. 2003. *Discovery learning about cocoa. An inspirational guide for training facilitators.* Egham, Surrey, UK: CABI Bioscience.

Walker, CA, and van West, P. 2007. Zoospore development in the oomycetes. *Fungal Biology* Reviews 21: 10–18.

Walters, D. 2011. *Plant defense: warding off attack by pathogens, herbivores and parasitic plants.* Oxford, UK: Wiley-Blackwell.

Walters, D. 2017. *Fortress plant: how to survive when everything wants to eat you.* Oxford, UK: Oxford University Press.

Walters, DR, Ratsep, J, and Havis, ND. 2013. Controlling crop diseases using induced resistance: challenges for the future. *Journal of Experimental Botany* 64: 1263–1280.

Wanger, TC, Darras, K, Bumrungsri, S, Tscharntke, T, and Klein, AM. 2015. Bat pest control contributes to food security in Thailand. *Biological Conservation* 171: 220–223.

Wanger, TC, Hölscher, D, Veldkamp, E, and Tscharntke, T. 2018. Cocoa production: monocultures are not the solution to climate adaptation—response to Abdulai et al. 2017. *Global Change Biology* 24: 561–562.

Way, MJ, and Khoo, KC. 1991. Colony dispersion and nesting habits of the ants, *Dolichoderus thoracicus* and *Oecophylla smaragdina* (Hymenoptera: Formicidae), in relation to their success as biological-control agents on cocoa. *Bulletin of Entomological Research* 81: 341–350.

Wessel, M, and Quist-Wessel, PMF. 2015. Cocoa production in West Africa: a review and analysis of recent developments. *NJAS—Wageningen Journal of Life Sciences* 74–75: 1–7.

What the Apothecary Ordered. n.d. Accessed April 20, 2018. https://apothecaryads.tumblr.com/post/96162203734/robert-gibson-and-sons-chocolate-worm-cakes.

Wheeler, AG. 2001. *Biology of the plant bugs* (Hemiptera: Miridae): *pests, predators, opportunities*. Ithaca, NY: Comstock Publishing/Cornell University Press.

Wielgoss, A, Tscharntke, T, Rumede, A, Fiala, B, Seidel, H, Shahabuddin, S, and Clough, Y. 2014. Interaction complexity matters: disentangling services and disservices of ant communities driving yield in tropical agroecosystems. *Proceedings of the Royal Society B* 281: 20132144. http://dx.doi.org/10.1098/rspb.2013.2144

Wilson, PK, and Hurst, WJ. 2012. *Chocolate as medicine: a quest over the centuries*. Cambridge, UK: Royal Society of Chemistry.

[WIPO] World Intellectual Property Organisation. 2017. "Breathing new life into Trinidad and Tobago's cocoa sector." https://www.wipo.int/wipo_magazine/en/2017/05/article_0005.html

Wong, J. 2017. *How to eat better*. London: Mitchell Beazley.

Wong, SY, and Lua, PL. 2011. Chocolate: food for moods. *Malaysian Journal of Nutrition* 17: 259–269.

Wood, GB, and Bache, F. 1834. *Dispensatory of the United States*. Philadelphia, USA: Gregg and Elliot. Cited by Wilson, PK, and Hurst, WJ. 2012. *Chocolate as medicine: a quest over the centuries*. Cambridge, UK: Royal Society of Chemistry, p 85.

Wood, GAR, and Lass, RA. 1985. *Cocoa*. Fourth Edition. Harlow, UK: Longman.

Worden, JR, Bloom, AA, Pandey, S, Jiang, Z, Worden, HM, Walker, TW, Houweling, S, and Rockmann, T. 2017. Reduced biomass burning emissions reconcile conflicting estimates of the post-2006 atmospheric methane budget. *Nature Communications* 8: article 2227. Doi: 10.1038/s41467-017-02246-0.

World Atlas. n.d. "Top 10 cocoa producing countries." Accessed 21 June 2018. https://www.worldatlas.com/articles/top-10-cocoa-producing-countries.html.

World Cocoa Foundation. n.d. "Cocoa & Forests Initiative." Accessed September 27, 2018. https://www.worldcocoafoundation.org/initiative/cocoa-forests-initiative/.

World Cocoa Foundation. n.d. "Climate smart cocoa." Accessed August 2, 2019. https://www.worldcocoafoundation.org/initiative/climate-smart-cocoa/.

Yanagawa, A, Yokohari, F, and Shimizu, S. 2008. Defence mechanism of the termite, *Coptotermes formosanus* Shiraki, to entomopathogenic fungi. *Journal of Invertebrate Pathology* 97: 165–170.

Yang, JY, Scascitelli, M, Motilal, LA, Sveinsson, S, Engels, JMM, Kane, NC, Dempewolf, H, Zhang, D, Maharaj, K, and Cronk, QCB. 2013. Complex origin of Trinitario-type *Theobroma cacao* (Malvaceae) from Trinidad and Tobago revealed using plastid genomics. *Tree Genetics and Genomes* 9: 829–840.

Yen, JDL, Waters, EK, and Hamilton, AJ. 2010. Cocoa pod borer (*Conopomorpha cramerella* Snellen) in Papua New Guinea: biosecurity models for New Ireland and the autonomous region of Bougainville. *Risk Analysis* 30: 293–309.

Young, AM. 2007. *The chocolate tree. A natural history of cacao*. Gainesville, USA: University Press of Florida.

Yuan, S, Li, X, Jin, Y, and Lu, J. 2017. Chocolate consumption and risk of coronary heart disease, stroke, and diabetes: a meta-analysis of prospective studies. *Nutrients* 9: 688.

Zarrillo, S, Gaikwad, N, Lanaud, C, Powis, T, Viot, C, Lesur, I, Fouet, O, Argout, X,

Guichoux, E, Salin, F, Loor, R, and others. 2018. The use and domestication of *Theobroma cacao* during the mid-Holocene in the upper Amazon. *Nature Ecology and Evolution* 2: 1879–1888.

Zhang, A, Kuang, LF, Maisin, N, Karumuru, B, Hall, DR, Virdiana, I, Lambert, S, Purung, HB, Wang, S, and Hebbar, P. 2008. Activity evaluation of cocoa pod borer sex pheromone in cacao fields. *Environmental Entomology* 37: 719–724.

Zhang, D, Martínez, WJ, Johnson, ES, Somarriba, E, Phillips-Mora, W, Astorga, C, Mischke, S, and Meinhardt, LW. 2012. Genetic diversity and spatial structure in a new distinct *Theobroma cacao* L. population in Bolivia. *Genetic Resources and Crop Evolution* 59: 239–252.

Zhang, D, and Motilal, L. 2016. "Origin, dispersal, and current global distribution of cacao genetic diversity." In *Cacao diseases: a history of old enemies and new encounters*, edited by BA Bailey and LW Meinhardt, 3–31. Switzerland: Springer International.

Zhang, Y, Massel, K, Godwin, ID, and Gao, C. 2018. Applications and potential of genome editing in crop improvement. *Genome Biology* 19: 210. https://doi.org/10.1186/s13059-018-1586-y

Index

DALE WALTERS is emeritus professor of plant pathology at Scotland's Rural College (SRUC). Until his retirement in February 2016, he was leader of the Crop Protection team at SRUC's campus in Edinburgh, where his research focused on induced resistance to plant pathogens. He obtained his BSc from Wye College, University of London, in 1978, and his PhD from Lancaster University in 1981, where he was also awarded a DSc in 1999 for his research on plant-pathogen interactions. He is the author of more than 160 scientific papers and has edited three books and written two textbooks, *Plant Defense: Warding off Attack by Pathogens, Pests, and Parasitic Plants* and *Physiological Responses of Plants to Attack*. His most recent book, aimed at a general audience, was *Fortress Plant: How to Survive When Everything Wants to Eat You*. His work on induced resistance to plant pathogens has been covered in the media.

Printed in the United States
By Bookmasters